REGULATORY POLITICS
AND ELECTRIC UTILITIES

REGULATORY POLITICS AND ELECTRIC UTILITIES

A Case Study in Political Economy

DOUGLAS D. ANDERSON
Graduate School of Business Administration
Harvard University

 Auburn House Publishing Company
Boston, Massachusetts

Library of Congress Cataloging in Publication Data

Anderson, Douglas D.   1949–
   Regulatory politics and electric utilities.

   Includes index.
   1.   Electric utilities—United States.   I.   Title.
HD9685.U5A7   363.6′2   80-26943
ISBN 0-86569-058-8

Printed in the United States of America.

For my parents—

Desmond L. Anderson and Loila F. Anderson

PREFACE

During the past decade, the lives of few business leaders have been more harried than those of executives in the electric power industry. Massive changes in the economic, technological, and political context of producing electricity have led to organized public attacks on the way in which the industry typically does business. Very little of what the industry does, from the choice of power generation technologies to the financing of new plants, has escaped the attention of outside political activists. Nowhere is the influence of outsiders more keenly felt than on the touchy subject of how the industry structures its prices. Critics have charged that some practices once widely accepted, such as the declining block rate structure (the method whereby utilities charge customers less per kilowatt-hour of electricity with each additional block of consumption), are inefficient or unfair. In some states utilities have actually been forced to abandon traditional methods; in every state regulatory and utility executives have had to grapple with thorny intellectual and political questions having to do with the way in which the industry produces and markets its output.

The lives of public officials responsible for regulating the industry's affairs—especially the state utility commissioners—have been no less hectic or confusing. Caught between the demands of utilities to raise rates, of consumers to keep rates down, and of environmentalists and others to "do something" about conservation and the energy crisis, commissions have been asked to mediate some of the most rancorous of recent domestic political disputes and to take on planning and pricing tasks historically unfamiliar to them.

It has been difficult to find patterns in the flurry that has enveloped state regulatory commissions. Theory has not helped. Conditions vary from state to state, but as a general proposition, the idea that state utility commissions are in the "pocket" of industry—the inherited wisdom—has been woefully inadequate for explaining

business-government relations in the largest states. In some instances what the electric utilities have wanted has been decisive; in
other instances it has not. (For that matter, utilities have not always
agreed on what government should do.) But the matter is even more
complex than that. At times, knowledge about a regulator's personal
goals and orientation has been sufficient information with which to
predict public policy, even in the face of widespread opposition from
industry, its largest customers, and the regulator's own staff. At
other times, neither the motives of regulators nor the wishes of
utilities can account for the outcome of the regulatory process.

This complexity is troublesome to anyone who would like to make
sense out of business and government interaction. It is disconcert-
ing, untidy, and confusing not to have a simple, universally applicable rule that predicts who will win the regulatory game and
influence the way in which it is played. During the 1950's and 1960's
most observers contented themselves with the notion (based on
fashionable political theory) that industry groups (electric utilities in
this case) would end up sooner or later dominating the process that
was supposed to regulate them. Whatever its merits as a predictor of
agency-utility relations in those two decades, that view is now simply out of date.

This book grew out of an attempt to take a fresh look at regulatory
behavior by focusing on the state regulation of electric utilities from
1968 to 1978. The questions that I asked myself were the following:
Why did it appear during this period that electric utilities were
sometimes successful in obtaining the type of public policy they
wanted and sometimes not? Why did regulatory executives try at
times to muffle the controversy surrounding their decisions and at
other times go out of their way to provoke controversy? Why did
regulators sometimes behave as if they were political entrepreneurs
and sometimes behave much more like bureaucratic managers?
Why could the staff of a regulatory agency exert an independent,
controlling influence over the content and process of regulation in
some cases, and in other cases discover that it was totally incapable
of affecting the regulatory outcome?

As I thought about these several questions, I concluded that the
answers (let alone the questions themselves) were not to be found in
conventional theories of regulation, and that a new conceptual
framework was needed. Chapter 1 consists of my effort to develop
such a framework, one that focuses attention on the terms of trade

that exist between a regulatory agency and the client groups in its environment, and on the internal dynamics of regulatory bureaucracy. I believe the framework to be generally useful in analyzing the politics of regulation, but in this book it is employed only in evaluating utility regulation by the states. In particular, I have concentrated on recent efforts by state commissions to reform the rate structures of electric utilities.

If the conceptual framework stands at the head of this book, then the body of the book consists of two historical chapters and two case studies. The first historical chapter (Chapter 2) is devoted to an analysis of the origins of state utility regulation, and is motivated by the assumption that forces present at the creation of a regulatory agency are important for understanding the nature of tasks assigned to that agency. The second historical chapter (Chapter 3) takes a look at more recent structural changes in the process of electric utility regulation. These structural changes have resulted in new demands that regulators take an active role in designing electric utility rate structures. Chapters 4 and 5 investigate these demands and the regulatory response of two large state regulatory bureaucracies, the New York Public Service Commission and the California Public Utilities Commission. The final chapter explains why these two state commissions selected the particular (and different) approaches they did and draws general conclusions about the nature of regulatory politics.

The second reason I undertook research into the politics of utility regulation is suggested in the subtitle of this book: "A Case Study in Political Economy." While the term *political economy* means different things to different people, I use it to describe inquiry into the politics of economic policy. One observation that can be made about economic regulation in the 1970's generally, I think, is that during the decade the advice of economists began to influence regulatory choices in a way that even leading members of the economics profession could not have foreseen ten years earlier. This is especially true in the case of electric utility rate structure regulation. It was during the 1970's that the principles of marginal cost were first applied by the industry (under the direction of state commissions) in pricing electricity. Why did it take so long for economic wisdom to be accepted by regulators and business executives in this sector of the economy? What changed that made theoretical principles not only acceptable—politically—but expedient?

My thinking about the political economy of regulation has benefited greatly from the influence of two masterful teachers, and it is with great pleasure that I acknowledge my debt to them. James Q. Wilson was the first to introduce me to the analysis of bureaucratic behavior. Coursework and discussions that I was privileged to have with him during four years of graduate study were the source of most of the concepts developed in this work. His influence at each step of this project has been profound. Richard E. Caves, more than any other person, taught me about the economics of regulation. He brought sharp and candid criticism to the reading of numerous drafts of this study, along with a good deal of warmth and wit. He has helped me enormously by constantly challenging me to dispel his doubts.

Many other individuals have helped me to gain a better understanding of the regulatory process and have read and commented on portions of this study. I am especially grateful to Stephen Breyer, Christopher DeMuth, Arnold Howitt, Paul Joskow, Marc Landy, Robert B. Reich, Robert Smiley, and Richard Zeckhauser; to my colleagues at the Harvard Business School, Joseph L. Bower, M. Colyer Crum, Robert A. Leone, George C. Lodge, Thomas K. McCraw, Gale D. Merseth, Michael E. Porter, Richard H. K. Vietor, and Louis T. Wells; and to members of the faculties of the School of Public Policy and the Department of Political Science at the University of California (Berkeley), the Graduate School of Business and Public Administration at Cornell University, and the Institute of Policy Sciences and Public Affairs at Duke University who commented on earlier drafts of the book's conceptual framework.

I also owe a great intellectual debt to many individuals whom I interviewed at the New York State Public Service Commission and the California Public Utilities Commission and in the numerous other governmental and private organizations that form the environment of those two agencies. These interviews were conducted during the course of two summers in 1976 and 1977 and on several occasions during 1978. My informants have been generous with their time and with their trust. All of the quotations that appear in this book, unless otherwise specified, are taken from these interviews. Many of those whom I interviewed have read and commented on drafts of the case studies and corrected factual errors. I am, of course, responsible for any errors—whether of fact, of interpretation, or of design—that remain.

I am also happy to acknowledge the financial support of the Alfred P. Sloan Foundation through its grant to Harvard University to promote the study of public management; without such assistance, it would have been impossible for me to have undertaken the field research in California and New York. The Division of Research at the Harvard Business School provided the support that enabled this book to be rewritten for the final time, for which I am very grateful. Although written for this book, portions of each of the chapters except Chapter 1 have appeared previously in James Q. Wilson, ed., *The Politics of Regulation* (New York: Basic Books, 1980) and are reproduced by permission here. Mrs. Shelly Reed cheerfully typed many of the drafts and suggested editorial corrections.

My greatest debts are personal, however, and owed to my family: to my parents for a lifetime of inspiration; to my wife, Elaine, for giving so selflessly throughout the entire period that this book was in draft; and to my children, Amy and Elliot, and now Emily, for providing the comic relief necessary to avoid taking it all too seriously.

D.D.A.

CONTENTS

Chapter 1

THE POLITICS OF REGULATION

The public view of independent regulatory commissions has undergone a curious evolution. Back in the heroic days when these commissions were being set up, there was considerable hope that government regulation would serve to correct failures of the market mechanism and thereby promote economic efficiency. Beginning with the establishment of the Interstate Commerce Commission in 1887, however, these hopes have been subject to frequent disappointment, to the point that in recent decades it has become the commonly held view that regulation by its very design promotes not the public interest but only the private interests of the companies being regulated. The public now typically thinks of government regulatory commissions as lethargic, inept, complacent, and overly eager to please industry—in a word, "captured" by the very industries they are supposed to regulate.

The capture theory has received considerable support from the academic community. Indeed, the vast majority of scholarly articles on regulation in the recent past have been devoted to debunking the earlier "public interest" view of regulation. However, the "capture" theory, which deep down is just the "public interest" theory turned on its head, fits reality little better than its predecessor. Why, for instance, would a "captured" Civil Aeronautics Board go against the wishes of most airlines and proceed with a program of deregulation? If the Federal Power Commission has been in the pocket of the natural gas producers for all this time, why have the producers had

1

such trouble obtaining deregulation of interstate natural gas prices? Does the Securities and Exchange Commission really behave as an industry captive? On a somewhat broader level, how does the capture theory explain the failure of the powerful automobile industry to fend off costly safety regulations? And how does the capture theory explain the plethora of environmental regulations that have been adopted in recent years?

While the capture theory may provide a fair description in some cases, it fails to explain the behavior of regulatory agencies in too many instances to warrant the wide acceptance it has received. It is the contention here that industry capture is a special case that should be viewed in the context of a more comprehensive theory— one that treats regulatory agencies as organizations that are both political and bureaucratic.

As political organizations, regulatory agencies have constituencies in the society at large. Clearly, the companies in the regulated industry form a significant client group, but there are others. Since the 1960's, public interest lobbying groups have loomed large in particular areas. In general, a regulatory agency will be tightly or loosely constrained by its client groups in accordance with its need for external support and the availability of that support. If an agency needs support badly and there is little available, it will have to pay a high price for support, and it will look like an agency that has been captured.

Agencies, however, are also bureaucratic entities, and as such they have to respond to the desires of an internal constituency, the bureaucrats who work in the agency. The power of this constituency varies from agency to agency. In some, the senior members of the professional staff may be seen as more powerful even than the commissioners.

Naturally, regulatory agencies vary in their responsiveness to outside political pressures and to internal bureaucratic imperatives. We can, however, often determine in advance when an agency will primarily be moved by politics and when bureaucratic considerations will be preeminent. Bureaucracy is most important in agencies when decisions grow out of extensive study of technical issues and where the work of the staff is therefore of great importance to the outcome. On the other hand, when decisions are essentially questions of policy rather than the development of technical findings, the role of external constituencies looms larger.

Examining the Capture Theory

Before we look at these ideas in greater detail, though, let us examine the capture theory more closely. Actually, it may be too generous to accord the status of theory to the collection of writings that undergird the conclusion that regulatory agencies act as if they were the servants and protectors of industry. Nevertheless, most discussions refer to the "capture theory," so the term will be used here as well. The theory is supported by two distinct schools. The older school, which may be called the "natural life cycle" school, has been propounded mainly by sociologists, legal scholars, and political scientists. The second and newer school has been developed primarily by economists; it may be called the "economic" school.

Two Schools of Thought

The natural life cycle school uses the analogy of youth and old age to explain how agencies set up to guard the public interest can wind up, over time, guarding corporate interests instead. This analysis holds that commissions are born only when public outrage over unfair business practices reaches a peak. At that point, a group of otherwise disparate interests forms a coalition in support of regulation. The industry to be regulated is, however, almost by definition a powerful one, and therefore the final legislative mandate is normally the result of extensive compromise. The legislators are likely to urge, in vague terms, that the new commission be "fair and reasonable" and develop such regulations as "the public interest, convenience and necessity" may dictate.

The compromises of the legislative battle are only a first step for the newly regulated industry, which may be seen less as defeated than aroused. On the other hand, the victorious pro-regulatory coalition is likely to disintegrate upon the establishment of the watchdog agency. Public outrage is short-lived; the agency's deliberations are technical; interest groups that once united to support the idea of regulation now divide over the details of regulation. As apathy replaces fervor, the legislature and executive withdraw support, both political and financial.

Isolated politically, the agency yields up the ambitious goals of its youth and seeks to strike a bargain with its industry. As special interests gain more and more control over substantive decision-making, the brighter and more creative members of the staff desert the agency rather than remain and be captured. Gradually but inexorably the infirmities of old age set in, and surrender is completed. Only the inept, or those who come from industry and are likely to return to it, are left to carry out the task of regulation.[1]

The "economic" school reaches similar conclusions but uses analogies from the marketplace rather than from biology.[2] Within the economic school there are two basic approaches. One approach is to apply the theory of consumer behavior to the behavior of regulatory commissioners. Key elements of that theory, of course, are the motivational assumptions of self-interest and rationality and the analytical technique of maximization subject to constraint. According to this theory, regulators seek to maximize their own utility. Unfortunately, the nature of the regulatory system gives regulators few incentives to put the general interest before special interests. Commissioners rarely make a career out of regulatory service. Typically, the terms of commissioners are short, and the probability of reappointment is slight since appointments often overlap presidential or gubernatorial terms. However committed to regulation a commissioner may be, he must give some thought to what he will do when he leaves the agency. Command of the technical details of transportation rate schedules or of some other equally arcane body of information is not a generally marketable product. The one place commissioners are sure to be able to cash in on their expertise and personal contacts is in the industries they regulate. A cantankerous commissioner—one who is perceived as antibusiness—is not likely to find employment in the industry he regulated once his term is completed. Even commissioners who plan to stay on with commissions must be concerned with business sentiment. Commissioners unfriendly to business are subject to constant attack in the press and in trade journals. They can be sure that if they make life too difficult for the industries they regulate, their reappointment will be bitterly opposed. To guarantee future employability, regulators seek to fashion policies that pacify industry. Commissioners take care of industry so that industry will take care of them.

Lee C. White, former chairman of the Federal Power Commission, recites the following parable on the "care and feeling" of regulators at the federal level:[3]

> *A successful lawyer in Keokuk is appointed by the President to serve on an independent regulatory agency or as an assistant secretary of an executive department that exercises regulatory functions. A round of*

*parties and neighborly acclaim surround the new appointee's depar-
ture from Keokuk. After the goodbyes, he arrives in Washington and
assumes his role as a regulator, believing that he is really a pretty
important guy. After all, he almost got elected to Congress from Iowa.
But after a few weeks in Washington, he realizes that nobody has ever
heard of him or cares much what he does—except one group of very
personable, reasonable, knowledgeable, delightful human beings who
recognize his true worth. These friendly fellows—all lawyers and officials of the special interests that the agency deals with—provide him
with information, views, and most important, love and affection. Except they bite hard when our regulator doesn't follow the light of their
wisdom. The cumulative effect is to turn his head a bit.*

The motives of commissioners and the incentives of the regulatory
system lead agency heads to engage in a great deal of what George
Hilton calls "minimal squawk" behavior. But the pacification of industry, however necessary, is not the only goal of commissioners. In
Hilton's view, "Regulatory bodies are established at least nominally
to protect the public, and the public interest is defined in their
procedures as a set of complaints heard in adversary proceedings.
Failure to deal with such complaints in a tolerable manner results in
hostile publicity and, more important, in adverse feedback to legislative bodies upon which the commissions depend for budgetary support, rectification of adverse court decisions, and ultimately for
preservation of the regulatory system. . . . Both for the maintenance
of the individual commissioner's reputation and for perpetuation of
the regulatory system which he is administering, such complaints
must be dealt with in a parallel process of *ad hoc* pacification." As a
result the "basic behavior of regulatory commissions" is to satisfy
both the demands of the regulated industry and those of its largest
customers by generating a monopoly gain in one activity and then
dissipating at least part of the gain in uneconomic activity. The
medium through which this process is carried out is the rate structure of the regulated industry, which regulators manipulate or allow
utility managers to control to assure that one group of customers
subsidizes another.[4]

The life of regulators is relatively unburdened as long as the system of cross-subsidies can be implemented. But that is not always
possible. The problem for regulators is that abnormally high returns
in one segment of the industry stimulate technological innovation
and encourage new firms to enter the industry. Successful innovation or entry by new firms leads to charges by the industry of "cream
skimming," as in the case of budget airlines that compete away prof-

its on high-density, long-haul routes without providing offsetting service to low-density, less-profitable routes. Thus, to preserve their power as protectors of industry, agencies seek to extend their jurisdiction and bring the new technologies and firms under their purview—if they cannot block innovation or the entry of new firms. One observer has called this tendency of regulation to spread the "tar-baby" effect.[5]

A Policy-Making Buyout by Industry

An alternative formulation of the economic theory of regulation stresses that policy-making is not captured but is purchased by industry. This is the path taken by George Stigler, among others, who maintains that "regulation is acquired by the industry and is designed and operated primarily for its benefit."[6] Instead of emphasizing the motives of commissioners, this theory develops a somewhat different model of the determinants of demand and supply for regulation, a service that consists of the exercise of the coercive power of the state. Consumers of regulation are economic groups such as industries and occupations which value the state's power for its usefulness in enhancing their economic status. The more effective government is in improving an industry's profitability, the greater will be the industry's demand for regulation. This means that government must contribute something to the market power an industry already possesses. Also, the more the benefits of regulation can be reserved for the group that successfully buys the service, the more willing groups will be to bid for the regulation. To the extent that regulation confers a public-good-like benefit—that is, regulation allows competitors to obtain some of the benefits of regulation without paying for them—demand will be diminished.

Regulation, in this model, is not costless. It is sold at a price by the government, although it is not quite clear what or who the government is. Stigler assumes the sellers are those who control and discipline politicians—in his view, the political parties. Parties have operating and maintenance expenses in addition to election expenses. Industrial groups obtain regulatory protection by contributing votes and other resources to party vendors.

The implications of the model are that large industries will be more successful than small industries or occupations in obtaining regulation because, with a fixed political market, the costs per unit of protection do not rise in proportion to industry size. Highly concentrated industries have more resources to buy regulatory protection because they can induce their members to contribute to such an effort more easily than less concentrated industries, but they have less need to buy protection because they already have significant market power. Entry or service restrictions will be adopted as policy instruments rather than as direct subsidies to avoid sharing benefits with competitors not part of the agreement.[7]

Reformulating the Theory of Government Regulation

The idea that regulatory policy-making will either be captured or bought by industry has such intuitive appeal for many persons that, when research provided some support for the notion, the argument over the sources and ends of regulation was more or less settled. The largest part of the literature on regulation in the recent past has been devoted to debunking the "public interest" view of regulation—namely, that regulation serves, and was created for, the purpose of correcting market failure and promoting economic efficiency. It is now a commonly held view that regulation by its very design protects only producers. If regulators acting in their own self-interest do not assure a capture-type result, large private groups, acting to cartelize their industry and to enhance their economic status, will. The strongest evidence in support of the producer protection hypothesis comes from the regulation of three oligopolistic or competitive industries: interstate airlines, railroads, and freight motor carriers. This evidence is familiar, has received an extensive reading, and will therefore not be reviewed here.[8]

In the rush to reject the public interest view of regulation in favor of the producer protection view, evidence contradicting—or at least casting doubt on—the hypothesis has been conveniently overlooked. But instances where regulatory policies have not benefited large industrial groups or—what is even more striking—where policies have conferred substantial costs on industry are too numerous to ignore.[9] Furthermore, the capture theories cannot account for changes in agency policy that are not increasingly beneficial to industry. Indeed, in a reversal of the life cycle school, commissions that regulate electric utilities may be seen as being "old and captured" early in their existence, but much more youthfully independent in recent years. It is possible that these transformations from old age to youth characterize the regulation of other industries as well. A related problem is the capture theory's failure to explain why a regulatory process that once benefited industry remains unaltered when changing economic conditions render it harmful to industrial interests.[10]

These limitations encourage a reformulation of the theory of government regulation in a pluralistic, free enterprise system. Such a reformulation has two tasks. The first is to shed light on the conditions that result in agency capture and those that preserve agency autonomy. The second is to refine our understanding of the determinants of regulatory behavior when agencies are not captured. The remainder of this chapter will be devoted to these two tasks.

External Pressures: The Political Economy of Regulation

If we think of the agency as a firm in a factor market, bidding for inputs in order to maintain itself and survive, some of the limitations of the producer protection theories can be overcome. Some people may immediately object that government agencies, unlike firms, do not face the threat of extinction. The Post Office, it is argued, has little fear of extinction no matter how badly it blunders. It continues to plod along because the service must be performed and because Congress is not likely to allow it to be taken over by a competitor for fear of incurring the wrath of postal employees and their families. Regulatory agencies are not like that, however. Even if they are not abolished, they can be reorganized. Their powers can be formally transferred elsewhere or, in the give and take of politics, another agency may simply move into a regulatory vacuum. The manager of a regulatory agency will therefore behave like the manager of a firm in that he will seek his agency's maintenance and survival as a goal rather than consider it a given.

A regulatory agency is, of course, not entirely like a firm. One of the major differences is that agencies frequently do not set their own tasks but receive them as a mandate; likewise, their ability to marshal resources to perform a particular task is limited by external directives and external control over purse strings. These limitations constrain an agency's capacity for innovation. There is another constraint which is perhaps less obvious. Businessmen more or less automatically limit the firm's tasks to an agenda consistent with the firm's resources. Agencies, on the other hand, can seldom discard tasks on their own initiative. They therefore frequently find themselves spread much too thin. It is this fundamental difference in the ability to accept or reject tasks that leads Joseph L. Bower to describe business strategy as the "application of massive resources to limited objectives" and the strategy of a governmental agency, by contrast, as the "application of limited resources to massive objectives."[11] Firms and regulatory agencies are alike, however, in their desire to manage their resources in ways that are both efficient (in that they do not waste resources) and effective (in that they accomplish the tasks set out).

The resources of a regulatory agency are of two basic kinds: internal and external. The internal resources are the technical expertise of its staff and its legal authority as granted by the legislature and interpreted by the courts. The external resources of an agency are those inputs that come from its clients, who may be broadly defined as any individual or group whose status is affected by what the agency does.

It is not the absolute level or amount of resources available to an agency executive that will determine his or her behavior but rather the agency's resources relative to its commitments and obligations. Changes in the technology or economic environment of a regulated firm can cause an agency's obligations to rise faster than its resources. In the late 1960's, for example, inflation and the slowing of technological improvements in the electric power industry caused the number of rate cases brought before state public utility commissions to rise dramatically, with little or no offsetting increase in the internal resources available to most of these commissions to deal with the additional workload (see Chapter 3). Resources relative to commitments fell, making it more difficult for commission executives to maintain their organizations and encouraging them to look outside the agency for help in coping with their new burdens. An increase in an agency's legislatively determined obligations could have a similar effect; so could a reduction in any of its internal resources, obligations held constant. The latter phenomenon could happen either because the legislature or the courts decide to restrict an agency's authority or cut its budget, or because its technical staff through loss of morale withdraws efforts or resigns. Even under normal circumstances, internal resources are unlikely to be sufficient for agency maintenance and survival, which leads to the following proposition: When an agency is unable to maintain itself and accomplish its task with internal resources alone, it will seek support from external sources.

The Demand for External Support

There are four elements that help to determine the extent of an agency's demand for external support: (1) its legislative mandate, (2) the complexity of its tasks, (3) the tasks' controversiality, and (4) the visibility of regulatory outcomes. First, what is the nature of the agency's legal mandate? Do the laws and court opinions that form the basis of the agency's legal grant of power clearly specify a set of tasks or is the mandate vague? Is the agency asked to serve competing or ambiguous goals? An imprecise mandate or one that requires an agency to serve conflicting ends can cause the agency to become embroiled in controversy with a regulated industry, with the effect that many of its resources are devoted simply to fighting jurisdictional or definitional battles. Over time this situation can be not only costly but also demoralizing to the agency's staff. This result, in turn, may reduce the quality or quantity of such resources and make the agency more dependent on external sources. For years the Supreme Court's *Smyth v. Ames* decision had such an effect on regulatory commissions by entangling them in disputes over the proper valuation of a utility's rate base.[12]

Second, is the task technologically complex? Does it require the agency to engage in a great deal of analysis, forecasting, risk appraisal, or evaluation? A simple, highly operational task does not require much help from outside the agency. If, for example, the task of an agency is to regulate crop production and the problem is overproduction, it is a relatively simple matter to mail checks to farmers who pull acreage out of cultivation. If, on the other hand, the agency is charged with regulating a highly complex production process, it may have to rely heavily on outside sources to make its task manageable. In general, the more complex the task, the greater will be the agency's demand for client support.

Third, is the task controversial? Serious controversy is detrimental to the organization. It may be sparked by disagreement over the goals or policies of regulation, and it can be provoked outside or inside the agency. The more the people in an agency have a shared sense of mission, and the more they are devoted to particular methodologies for carrying out that mission, the less the likelihood of internal controversy. A common professional training, in law or engineering, for instance, can provide staff members with shared behavioral norms.

In another context, Alan Altshuler has noted the political advantage that accrues to an agency that has a "clear sense of direction and honest conviction" derived from a common professional outlook. "It was a crucial advantage of the highway engineers in freeway disputes," he writes, "that they possessed and believed implicitly in a set of clear normative propositions, applicable to the most important of their problems and convincing to the vast majority of people with whom they had to deal."[13]

Shared professional norms are not sufficient, of course, to eliminate disputes over an agency's task. Conflicts might arise from outside an agency because the agency is perceived as conferring highly concentrated benefits or costs or because of the existence of a competing governmental entity with a different organizational ethos.

Fourth and finally, how visible are regulatory outcomes? Controversiality and visibility are, of course, related. We often think that the more controversial are the tasks of a regulatory agency, the more visible will be its outcomes. But this is not always so. If a regulatory agency bestows at the same time concentrated benefits but widely distributed costs, it may be engaging in controversial but not very visible behavior. Consider, for example, the case of a public utilities commission that allows a regulated electric utility with an urban service area to merge with a number of smaller utilities in outlying areas. An argument could be made that the purpose of the merger was to facilitate greater efficiency through a larger scale of operations. Suppose, however, that the "real" reasons the utility commission favored the move was because it allowed the urban utility to spread its high property taxes over a larger customer base. If this policy had the intended effect of significantly reducing the rates paid by city dwellers and only marginally increasing the rates of rural users, it would be no less controversial for being relatively invisible.

In most instances, controversy can be expected to increase an agency's dependency upon a supportive external constituency. Internal squabblings among different professional groups within a regulatory staff or between the staff and the regulatory executives are likely to weaken the authority of the agency. They also increase the probability that outsiders, particularly the legislature, will become involved in the agency's affairs. Jurisdictional or policy disputes among governmental agencies are likely to have a similar effect. Some presidents, such as Franklin D. Roosevelt, have operated on the assumption that competition among agencies fosters bureaucratic creativity and vitality. But it is just as likely that an agency faced by a threat to its survival will devote its resources to bureaucratic in-fighting and that it will enlist its clients in the effort. If the controversy is provoked by a major crisis—a black-out in the case of electric utilities, bankruptcy of a major carrier, dramatic loss of life, catastrophic injury, or political scandal—the agency's survival will be threatened because otherwise supportive or indifferent clients will turn against it.

The role of visibility in agency-client group relations is even more complex. Certainly visibility increases the likelihood of opposition and therefore the need for external support; but visibility also tends to decrease the viability of a truly intimate relationship with any outside supporting group. As a result, one would expect that the more visible an agency's policy outcomes, the less will be its ability to arrange a "contract" for support with any particular client group.

The Resources Provided by an External Constituency

The nature of an agency's legislative mandate, the complexity and controversiality of its tasks, and the visibility of its outcomes are all elements bearing on agency demand for client support in maintaining itself. What are the resources that an external constituency can provide to a regulatory agency? There are basically three different types of resources, which may be classified as informational, political, and polemical.

Regulators use information to make decisions. They also use it as a basis for legitimacy. Because they are not elected, their democratic legitimacy is indirect; they therefore tend to seek technocratic legitimacy to bolster their position. External groups that supply good information help the agency not only with the decision-making task at hand but also with the agency's underlying legitimacy.

Regulated enterprises tend to have a natural advantage over other external actors in the provision of information because there are few people, even within agencies, who possess as much detailed information on an industry as those who work for companies in that industry. The more complex a certain production process, the greater is the informational advantage of those who operate it. This does not mean, however, that consumer or environmental groups cannot obtain and supply informational resources. They can and do, and in some instances this information is decisive. For example, in Chicago a 1970 canvass of coal suppliers by the Clean Air Coordinating Committee succeeded in locating ready supplies of low-sulfur coal and thereby supplemented the expertise of the Chicago Department of Environmental Control. This information was used to deny variances to industrial users who claimed they were unable to meet air quality standards for lack of adequate supplies of high-quality coal.[14]

The political resources of client groups consist of the influence they may have over governmental leaders who are capable of augmenting or reducing an agency's technical and legal resources or of changing its executives. There are numerous familiar ways whereby business and consumer groups strive to obtain political influence.[15] It is tempting to suggest that, from the perspective of a regulatory agency, the crucial point is not how clients get influence but how they use it. Business clients who have captured agencies have done so largely because they have used their political influence to deny agencies needed technical expertise, to challenge the agency's legal authority, or to threaten an executive's tenure. How they achieved this influence does not seem nearly so important as how they exercised it.

In the case of consumer and environmental groups, however, the obtaining of political influence may be as important as its use in that these groups often wield power indirectly through the media rather than acting directly upon political leaders. This fact is significant because, unlike discreet suggestions made in agency corridors, media publicity has the effect of changing latent controversies into highly visible ones. Such visibility, in turn, can have the possible effects of preventing a "contract" between an agency and a regulated business, strengthening agency resolve to adopt policies contrary to the wishes of regulated companies, or of causing business to modify future requests.

In Los Angeles, for example, a group called CAUSE (Campaign Against Utility Service Exploitation) successfully dramatized Pacific Telephone's 1976 proposal to charge 20 cents for information calls by encouraging telephone subscribers to request extra phone books. The organization adopted the view that, if people could not call information free of charge, they would need to have copies of all 13 phone books covering the 213 area codes. (Telephone companies traditionally distribute these books at no additional charge.) Even if

one relies on CAUSE's estimates that "thousands of people" requested extra phone books, that in itself was not likely to have
affected the Public Utilities Commission's ultimate denial of Pacific
Telephone's request. Of greater importance, certainly, was the
newspaper and television coverage that accompanied the campaign
and the group's final blow of returning many of the books to the front
door of the company's Los Angeles office after the decision was
announced. No one (other than CAUSE) would argue that the publicity was decisive in this case, but it did strengthen the commission's predilection to deny the request and may have affected the
company's strategy for future rate increases.[16]

The final resource offered agencies by client groups is polemical.
This type of support is highly useful to the agency, even though it
may be unintended. Groups seeking to influence a regulatory decision normally adopt an adversary posture in which they stake out
positions on opposite sides of a question. This leaves the middle
ground unoccupied, and the agency may then take its position there
with at least an appearance of reason and moderation.

In summary, agency demand for client support derives from its
need for external resources to maintain itself. This need, in turn, is
greater the more limited are an agency's internal resources relative
to its commitments. Legal mandates that are fuzzy and tasks that are
complex and controversial are likely to saddle agencies with "massive objectives" but only limited resources. Agency demand for external support will increase with the quality of the informational,
political, and polemical resources offered by client organizations.

The Determinants of the Supply of External Support

We do not have a fully satisfactory theory of the determinants of the
supply of external support. It is not difficult to explain the presence
of interest groups when the benefits or costs of regulatory policy are
concentrated. Under such conditions the costs of organizing can be
paid out of the returns from lobbying. As long as the group is relatively small, each individual's contribution (or failure to contribute)
can be noted and rewarded (or punished). More difficult is the task
of explaining organized activity when the costs or benefits of policy
are widely diffused. Under such conditions each individual knows
that his contribution will have a negligible impact on the group's
prospects and that he will likely go unnoticed if he fails to contrib-

ute. Since he will be able to enjoy any benefit the group achieves whether or not he participates, he has no incentive to join the group. Since everyone feels the same way, no cooperative behavior is possible.[17]

One of the primary reasons the producer protection hypothesis has acquired such credibility is that this view of interest group politics effectively limits the supply side of our model to industry and its largest customers.[18] But one of the most interesting developments in American politics since the 1960's has been the success of the consumer and environmental movements. It is possible that changes in communications—notably the advent of television—sustain these organizational efforts by creating new opportunities for political entrepreneurs to fashion loose "representative" interest groups whose membership is neither stable nor solicited for any contribution other than that of acting as an audience for the Nielsen ratings.[19] Access to television confers legitimacy upon the group's efforts to speak on behalf of a wide constituency and offers opportunities for public exposure and political advancement for its organizational elite.

Some theorists have suggested alternative scenarios for the advent of these "representative" interest groups. In one, the maintenance and enhancement needs of an existing organization induce it to expand its scope of activity to include new policy areas. This seems to have been the case with the Chicago Tuberculosis Institute, which became involved in the issue of air pollution control when air pollution began to be discussed in terms of its public health effects. Another theory suggests that regulatory agencies take it upon themselves to stimulate the organization of client groups—as though creating their own source of supply—by sharing technical expertise and encouraging citizen participation. This apparently was the strategy of the National Air Pollution Control Administration, the forerunner of the Environmental Protection Agency.[20]

Agency Autonomy as a Function of Resource Supply and Demand

It is not the purpose of this conceptualization to evaluate these accounts of how "representative" interest groups are formed, or to develop a more complete theory of the supply of external support. It is sufficient simply to observe that if an agency finds itself in need of any of these resources that outsiders can provide, it must bid for them. The price paid will be determined by the laws of supply and demand: Abundance of need will tend to drive the price up; so will scarcity of supply. In what coin is this price paid? The agency pays for its external support by sharing control over decision-making with its suppliers.

The scarcer the supply of external support, and the greater the demand for it, the higher will be the price that regulatory agencies will pay and the more they will behave like captured agencies. However, when an agency does not need to rely on client support to maintain itself or when it has multiple sources of supply for the external resources it requires, the price will be low and the agency's autonomy will be preserved.

This reformulation of the theory of regulation renders the notion of an agency "life cycle" unnecessary. Changing supply and demand conditions alter the degree of servitude or autonomy an agency exhibits without regard to "age." The new theory also frees us from the view that regulators will serve with single-minded devotion the interests of any one client group. A captured agency will share power with industrial groups as long as it has no alternative source for the external support it needs. If new sources of supply develop, the previously favored industrial group will see its grip over the agency begin to loosen. In an extreme case the initial dictators of agency policy lose all the control they once enjoyed (perhaps because the legitimacy of their political and informational resources becomes discredited). In that case, if the need for external support is still great, it is possible that the agency will be recaptured—but by a new coalition of interests.

There are limits to this model. Because we do not fully understand all the components of supply and demand, it is difficult to use the model to make a full explanation of various regulatory actions. The model is most useful in analyzing captured agencies because it corrects misconceptions that are built into the capture theory. There is a tendency to assume that a captured agency simply performs the will of the regulated industry, and that any deviations between the industry's position and the agency's position may be explained simply as anomalies. Such a conception is very attractive because it allows the analyst to ignore the agency. All the interesting questions pertaining to the choice and content of public policy can be addressed to the industry itself.

No agency, though, will ever behave in a way that corresponds precisely to the desires of its clients. Consider as an analogy the behavior of inmates in a maximum security prison. If ever there is true capture in our society, it is in a maximum security prison. Yet as Gresham Sykes pointed out two decades ago in his book, *The Society of Captives*, "Despite the guns and the surveillance, the searches and the precautions of the custodians, the actual behavior of the inmate population differs markedly from that which is called for by official commands and decrees."[21]

Clearly no agency is going to be as tightly constrained as a prison population. It would be ludicrous even to mention the analogy were it not for the wide circulation and loose acceptance that the capture theory has received. Even in instances of agency capture, it is unlikely that the agency will be so closely controlled by external forces as to be totally incapable of independent action. Almost always the agency will retain some space within which to deviate from the wishes of a favored client group. Even with that, agency capture should really be seen as a special case within a more general theory of regulation. Between the polar extremes of capture and recapture the agency preserves some freedom to develop and implement policies independent of its clients. This "zone of autonomy" will be wide or narrow, depending on the factors of demand and supply for external support.

Regulatory agencies are not all alike. Some may share a significant portion of their state-granted police powers with a single client group for a long period of time. Others, like the automobile with a nearly-flat front left tire, will have a tendency to favor policy outcomes which benefit a particular clientele but may be susceptible to a sharp counter-directional force applied to the steering mechanism. Still others, like the National Labor Relations Board, will display no permanent or stable bias in favor of one group, but will serve as the referees of an ongoing struggle between two or among more organized interests in which outcomes sometimes favor one faction and sometimes another.[22]

It is the interaction of supply and demand that allows us to understand this give and take, these shifts, and the apparent logical inconsistencies that go on in the regulatory environment. People are bargaining. People are negotiating. And they are doing so in a fluid environment.

As the zone of autonomy increases, however, the predictive value of the model declines. The behavior of uncaptured agencies is much more varied than that of captured agencies, and the model needs further refinement if it is going to be useful in explaining all this variety. We need to go back and look some more at the nature of the agencies themselves.

Bureaucratic Imperatives: The Behavior of Regulatory Officials and the Content of Regulatory Policy

In addition to behavioral differences, there are structural variations among regulatory agencies. Some are independent commissions chartered by Congress or state legislatures; others are bureaus within the executive branch of government. The earliest regulatory agencies were set up to control private economic behavior and performance. Obviously that is not the only kind of private behavior subject to regulatory control now. In recent years, agencies charged with developing rules to protect the environment and to promote health and safety have been among the most active. For definitional purposes, it is the regulatory function, and not an agency's structure or substantive orientation, that is of importance. A "regulatory agency" may therefore be defined as a public organization whose mission it is to make appraisals, develop rules, and issue policies for the control of private behavior. The most famous definition of "organization" is that of Chester I. Barnard: ". . . a system of consciously coordinated activities . . . or two or more persons."[23] A "bureaucratic" regulatory agency is an agency in which authority for managing the regulatory function or mission is divided among several appointed officials.

Earlier we saw that viewing the agency as a firm striving to maintain itself and survive was a useful perspective for discerning the conditions which are likely to lead to capture and those which preserve autonomy. Let us now develop these ideas further as a means for understanding the behavior of regulatory officials and the content of regulatory policy.

Point Decisions and Planning Tasks

Most regulatory tasks can be conveniently assigned to one of two categories. The first is basically a policy-setting task. The chief characteristic of such a task is that it involves what might be termed a "franchise," or "point decision"—that is, a single decision of the "yes-no," "locate here, not there" variety which allocates a scarce value to one of a number of actual or potential rivals. When the Federal Communications Commission decides to renew a certain television station's license, it is engaging in a point decision; so is the Civil Aeronautics Board when it approves a new route for one airline and not another. When a coal-fired electric generating plant is approved for construction at a particular site, it is because some one or several regulatory agencies have made point decisions.

The second type of regulatory task is that which involves a great deal of coordination, appraisal, or planning.[24] An example of a task of this sort would be a utility commission's efforts to process a request for a rate increase by a regulated company. Evaluating a state's energy needs, encouraging conservation, or promoting production are all tasks which require the coordination of a myriad of individual actions in appraising risks and developing plans. Clearly, most regulatory agencies have mixed responsibilities; some of their tasks are planning tasks and some call for point decisions.

The distinguishing characteristic of a planning task is that it requires agency executives to coordinate over time the behavior of a great many people who must be relied upon to make judgments and issue instructions in their absence and on their behalf. From the standpoint of an agency head, planning tasks present what may be termed a "bureaucratic" problem—namely, that of inducing people to behave precisely as the executive would were he or she in their place. The bureaucratic problem is especially acute if, in addition, the executive has no good way of ascertaining whether or not subordinates are following the agency's policies or aiding in accomplishing its goals.

A point decision, on the other hand, confronts agency executives with quite another type of problem. Its distinguishing characteristic is that it requires a decision on how (or to whom) to allocate a scarce resource. Once the decision has been made, there is no great need to induce cooperative behavior by the regulatory staff, and hence the need to control or direct staff behavior does not greatly constrain agency executives. Much more powerful constraints on the behavior of agency heads faced with point decisions are exerted by the conflicting external interests that are affected by the agency's actions.

It is the basic proposition of this study that when the task of an agency is one which requires much coordination, appraisal, or planning, the behavior of regulatory executives can be best understood in terms of their efforts to maintain the organization and enhance their position in it.[25] When, on the other hand, the agency's task is primarily one of choosing among discrete alternatives (making a point decision), the behavior of regulators can best be understood in terms of their attempts to forge a coalition of support among selected groups in the agency's environment.

Political and Managerial Roles of Agency Executives

Persons who occupy positions of major responsibility within regulatory agencies may be said to be in part political executives and in part managerial executives, depending on whether the primary influence over their behavior comes from outside or within the agencies. The reason for segmenting agency tasks is as follows: When the survival of an agency depends on its success in engaging in coordinated behavior, the agency executive will be evaluated on the basis of how well he produces and sustains cooperative effort, or, in other words, maintains the organization. James Q. Wilson points out that organizational maintenance "chiefly involves supplying tangible and intangible incentives to individuals in order that they will become, or remain, members and will perform certain tasks."[26] The need to conserve and enhance the supply of incentives by which the agency's staff is induced to perform their tasks powerfully constrains the actions of those responsible for maintaining the agency. Alternatively, when the need to induce cooperative behavior is not great either because the task is primarily one of weighing alternative values and choosing between them or because there is some unambiguous measure of the success of a particular policy, the executive will not be as constrained by the requirements of organizational maintenance and will be freer to respond to what he perceives to be the opportunities or risks in the external environment for his own or his agency's success.

Obviously, the behavior of executives even in agencies that are faced primarily with point decisions will not be totally unresponsive to the desires, beliefs, and cherished ideals of the agency's staff. One of the primary incentives that draws people to work for regulatory agencies is purposive: the desire to advance one's own conception of the public interest. Professional persons who staff agencies— engineers and attorneys, for example, in public utility commissions and medical doctors in the public health service—are powerfully influenced by the norms of their professions, another important source of incentives. Agency heads whose decisions consistently violate the staff's standards of medical or legal procedure or whose decisions regularly offend the ideological sensibilities of the staff may observe erosion in the quality of staff work or find significant numbers of staff members leaving the agency. Whether the decline in staff morale is a constraint on an agency executive will in large part be related to the executive's time horizon: How long does he expect to remain with the agency and, therefore, how likely is it that the agency's staff can successfully retaliate for unpopular decisions? Political executives whose tenure is short may be quite unconstrained by the requirements of organizational maintenance.

Similarly, managerial executives whose primary concern is maintaining the organization will be influenced by the external environment. Naturally they will be more constrained by a narrower zone of autonomy, but even with wide latitude it is unlikely their behavior will ignore environmental factors. It is possible that environmental factors—such as the lure of higher income in private industry or political controversies that bring into question the staff's commitment to the public interest will threaten the system of cooperative effort. When such a threat arises, the organization is subject to "strain," to use Wilson's concept. Strain may be induced by a serious conflict over the agency's purposes, the challenge of a rival organization, the loss of a sense of mission, or a decrease in the supply of incentives. Organizations cannot tolerate strain for long; indeed, it is a proposition here that managerial executives—but not necessarily political executives—produce and sustain cooperative effort by minimizing strain through some form of accommodation with their environment.

Four Typical Regulatory Situations

To explain fully the determinants of regulatory behavior, many factors would have to be considered. But the conceptual framework developed here makes it possible to characterize the regulatory process in terms of two major determinants: (1) the stringency of the agency's external constraints—is its zone of autonomy relatively wide or narrow? and (2) the nature of the agency's task—is the task a policy task involving a point decision, or is it a planning task requiring much cooperative effort? An agency's survival price may be high,

NATURE OF TASK

		Point Decision	*Planning Task*
	Tight	I. Capture	II. Conflict Minimizing
External Constraint			
	Loose	III. Entrepreneurial	IV. Bureaucratic

Figure 1.1   Four Modes of Regulatory Behavior.

low, or medium and regulatory tasks are often by nature mixed. Nevertheless, it is possible to formulate four ideal cases that are sufficiently discriminating to characterize most regulatory situations. These cases are presented in Figure 1.1. In each case the constraints on executive behavior are different, a situation that suggests variation in the consequent regulatory behavior.

Case I: The Capture Mode of Regulatory Behavior

The capture mode of regulatory behavior is a special case of the more general theory of regulation. In this situation executives face a tightly constrained point decision. The nature of the task at hand frees regulators from internal—that is, bureaucratic—constraints. At the same time, however, regulators are highly dependent on the support of an external client group. The wishes of the dominant client, whether it is the regulated industry or some other group, will be controlling. The views of the staff, to the extent that they differ from those of the client, will be ignored by regulators either because the staff's cooperation is not necessary to complete the task or because there exists some operational measure of the degree to which staff behavior complies with executive directives that enable regulators to keep the staff in line.

Because of the binding nature of the external constraint, regulators are incapable of exerting much independent will in this situation. The autonomy zone is narrow because the survival of the agency and a regulator's own future depend on his ability to please a single outside audience having a unitary point of view. The analyst who is able to decipher the client's agenda can, in this case, predict the outcome of regulatory policy.

It is unlikely, however, that an analyst will find many situations that conform to the rigorous specification of the capture mode. In a pluralistic society, regulators will seldom be completely controlled by the views of a single client group.[27] But even if regulators should be captured by an industry group, for example, this is no guarantee that they will be issued a single set of marching orders. An industry is not a monolith. It is made up of individual firms that face differing cost and competitive conditions and that therefore have different stakes in the regulatory process and different views on regulatory policy. Regulators will be rarely so tightly constrained that they cannot choose which firms within an industry they will benefit.[28] Moreover, unless a regulator has a short time horizon, that is, unless he does not intend to stay on the job very long, his votes on point decisions will necessarily reflect to a certain extent the staff's thinking. At some point even regulators whose policy positions are highly constrained by client groups will need to rely on the cooperation of their staff. When this happens we enter the next mode of regulatory behavior.

Case II: The Conflict-Minimizing Mode

When regulatory executives are caught between potentially competing internal and external constraints, their behavior will be designed to minimize conflict. It is in describing this mode of regulatory behavior that George Hilton's colorful term "minimal squawk" has its best application. In this situation, however, regulators must seek to pacify not simply the industry or its critics but also the members of their own bureaucracy. Their need to buy external support from a particular clientele means that they will operate under a tight external constraint. But the fact that they are faced with planning tasks means that they will have to encourage and sustain the cooperative effort of their staff. This places regulators under an internal constraint as well. If the wishes of the client and those of the staff should diverge, regulators will seek to fashion a compromise that minimizes organizational strains but accommodates the dominating client group—the case of the automobile with the flat front tire. Regulatory policies will be biased in favor of particular clients but susceptible to change when strain is too great.

Case III: The Entrepreneurial Mode

A polar extreme from the conflict-minimizing mode is the situation in which regulators are relatively free from both external and internal constraints. The zone of autonomy within which regulatory policy is fashioned is wide because a multiplicity of client groups provides regulators with an abundant supply of external support relative to the agency's need for this support. Regulators are freed from binding political pressures. Correspondingly, regulators are also freed from bureaucratic imperatives because of the nature of point decisions.

In this case regulatory behavior is best understood by focusing on the personal, professional, and purposive motivations of the regulators themselves. Regulatory executives act as if they were policy entrepreneurs, responding to perceived risks and opportunities for advancing their own interests through building coalitions. The economic theories of regulation which are built upon the assumption of commission utility maximization have their best application here. These theories, based on the idea of an exchange economy, suggest that even if a regulator were not compelled to develop and implement policies favorable toward business, he would do so on his own accord in exchange for future employment within the industry or for a peaceful existence while he served out his term on the regulatory bench. What utility maximization theorists sometimes fail to recognize, however, is that there are markets other than the "industrial market" for the human capital which commissioners amass while on the bench. In particular there is the *political market*. Since this market has been so widely overlooked, an explanatory comment seems in order at this point.

Commissioners responding to the political market are intent upon appealing to classes of voters for whom the regulatory process bestows highly diffuse but significant benefits or costs. As political entrepreneurs they seek to develop reputations as "consumer" or "environmental" advocates through the decisions they make on regulatory policy. This reputation, in turn, may be useful to them in securing future appointive or elective governmental positions.

The political market has always been a part of the regulatory environment, its emergence and reemergence episodic. One of the first well-known political entrepreneurs to cater to this market was Huey Long, who began his political career as a consumer advocate on the Louisiana Railroad Commission that later became the Louisiana Public Service Commission.[29] We will not attempt to account fully for the factors which give rise to the political market in one period or cause it to go into relative decline in another, but its ebb and flow seem related to shifts in the distribution of costs and benefits conferred by the regulatory process, and perhaps as well to changes in the technology of communication.[30]

·Regulatory executives who respond to the political market are *not* likely to engage in minimal squawk behavior. On the contrary, they are likely to seek confrontations in order to dramatize their political orientations and to gain a wider audience. For this reason, we might just as well have chosen to call this the "conflict maximizing" or "maximal squawk" mode of regulatory behavior.

Of course, regulator-designed conflicts subject agencies to increased scrutiny and may produce considerable strain within the organization. Staff members may become demoralized over what they consider to be opportunistic and unprofessional behavior by the executive. To reiterate, whether or not this strain proves to be a major constraint on the entrepreneurial behavior of a regulatory executive depends on the degree to which his agency is faced by planning tasks as well as point decisions, and especially on the regulator's time horizon. Regulators whose time horizon is short or whose tasks are comprised principally of making point decisions will ignore this strain with impunity.

Case IV: The Bureaucratic Mode

In contrast to the other three modes, the behavior of the executive whose agency must carry out a loosely constrained planning task is best understood in terms of his attempt to respond to the bureaucratic imperatives associated with maintaining the agency. This derives from the fact that regulators face few external constraints in this case but depend significantly on the cooperation of their staff to help them formulate and implement policy. The nature of the planning task itself, involving as it does the concerted efforts of many people, is what imposes the chief determinants on the behavior of regulatory executives. Because the requirements of organizational maintenance are so important in situations of this sort, let us call this the "bureaucratic" mode of regulatory behavior. Organizational maintenance involves not only providing a continuous stream of incentives to those who contribute necessary resources, as was noted earlier, but also managing an effective system of communications.[31]

In private bureaucratic organizations, managing an effective system of communications involves chiefly defining organizational positions—what Barnard calls the "scheme of organization"[32]—and matching those positions with trained and loyal personnel. The combination of personnel and positions describes the formal organization. Managing the formal organization is a primary function of the private business executive. In any organization, of course, the formal organization is not the only mechanism whereby cooperative efforts are secured. Of great importance is the "informal organization," consisting of personal contacts and interactions of people which occur without any specific conscious joint purpose and which are not governed by the formal organization.[33]

The dynamics of the informal organization as well as those of the formal organization are important for understanding executive behavior. In public organizations the informal organization is especially important because the public executive, unlike the private executive, has relatively little control over the structure of the organization or the assignment of personnel who occupy key positions. These elements are largely determined by the legislature and the requirements of the civil service. Because it is so difficult for public executives to transfer personnel and nearly impossible to use monetary incentives to induce loyalty and cooperative behavior, regulatory executives faced with planning tasks will rely extensively on the informal organization. A corollary of the proposition that executives seek to minimize organizational strain is the following: When the nature of the task is such that a great deal of appraisal, planning, or cooperation is required, the behavior of executives can be understood in terms of their failure or success in maintaining a system of informal contacts among those people whose efforts must be coordinated.

Conclusion

In this chapter conventional theories of government regulation suggesting that regulatory agencies are or eventually will be captured by large producers have been examined and found inadequate for the purpose of evaluating much of the regulatory experience of the past decade. To correct for these shortcomings an alternative conceptualization of the regulatory process was offered, proposing that regulatory agencies will be tightly or loosely constrained by client groups in accordance with the extent of their need for external support and its availability. We noted that the determinants of the demand for and supply of external support were subject to change, that changes in these factors have important ramifications for the zone of autonomy of the agency (how tightly the agency is constrained by external groups), and that these changes can be noted and evaluated. Next a series of propositions regarding the behavior of agencies in the absence of capture was offered that suggested the usefulness of discriminating between regulatory tasks that are "point decisions" and those that are planning tasks. We offered the general proposition that when the task of an agency is one requiring much coordination, appraisal, or planning, the behavior of its executives can best be understood in terms of their attempts to maintain the organization; but that when the agency's task involves a clear, simple choice among discrete alternatives, regulatory policy is more likely to be explainable in terms of executive response to perceived risks and opportunities in the external environment. Finally, these two major determinants—the degree of external constraint and the nature of the task—were combined to formulate four ideal cases thought to be generally useful in characterizing a wide spectrum of regulatory situations.

Now it is important to test the usefulness of this conceptualization by applying it to actual cases of government regulation. The cases chosen for analysis had to have three properties. First, they must allow us to evaluate structural changes in the process of regulation. Can we explain the tightening or loosening of external constraints on agency behavior in terms of our model of the supply and demand for agency support? What consequences do such structural changes have for the content of policy? Second, the cases must allow us to examine the many types of regulatory behavior that arise from different regulatory tasks and differing zones of agency autonomy. Finally, so that this conceptualization is not misapplied, the cases chosen must involve the regulatory behavior of public bureaucracies in which managerial authority is divided among several appointed officials.

The regulation of private electric utilities by New York and California qualifies on all three counts. The public utility commissions in these two states are among the largest of state regulatory agencies. Policies and procedures adopted by the two commissions in recent years provide examples of the "non-capture" ideal types developed above. In the past ten years, inflation and the advent of consumer and environmental groups have dramatically altered the external constraints faced by regulators, and this fact has had important consequences for the behavior of the two state agencies. In the chapters that follow we shall examine the origins of state utility regulation, and the structural evolution that has occurred in the political economy of electric utility regulation during the past decade, as well as the policy response to these developments by the New York and California commissions. In each case a comprehension of the changes in the nature of the tasks faced by utility commissions is important for understanding the regulatory response.

Endnotes

1. Marver Bernstein summarized and articulated the views of the natural life cycle school in *Regulating Business By Independent Commission* (Princeton: Princeton University Press, 1955).
2. *See* John R. Baldwin, *The Regulatory Agency and the Public Corporation* (Cambridge, Mass.: Ballinger, 1975); Ross D. Eckert, "The Los Angeles Taxi Monopoly: An Economic Inquiry," *Southern California Law Review*, 43 (Summer 1970), pp. 406–453; George W. Hilton, "The Basic Behavior of Regulatory Commissions," *American Economic Review Papers and Proceedings*, 62 (1972), pp. 47–54; Roger Noll, "The Economics and Politics of Regulation," *Virginia Law Review*, 57 (1971), pp. 1016–1032; Richard A. Posner, "Taxation by Regulation," *Bell Journal of Economics*, 2 (1971), pp. 22–50; George Stigler, "The Theory of Economic Regulation," *Bell Journal of Economics*, 2 (1971), pp. 3–21; Sam Peltzman, "Toward a More General Theory of Regulation," *Journal of Law and Economics*, 19 (August 1976), pp. 211–240. For reviews of the literature on regulatory agencies, see Paul L. Joskow, "Regulatory Activities by Government Agencies," *Massachusetts Institute of Technology Economic Department Working Paper No. 171* (December 1975), and Thomas K. McCraw, "Regulation in America: A Review Article," *Business History Review*, Vol. XLIX, No. 2 (Summer 1975), pp. 159–183.
3. Quoted in Robert R. Leflar and Martin H. Rogol, "Consumer Participation in the Regulation of Public Utilities: A Model Act," *Harvard Journal of Legislation*, 13 (1976), p. 242.
4. Hilton, "Basic Behavior of Regulatory Commissions," p. 48.
5. J. W. McKie, "Regulation and the Free Market: The Problem of Boundaries," *Bell Journal of Economics*, 1 (1970), pp. 6–26.
6. Stigler, "Theory of Economic Regulation," p. 3.
7. On this point John Baldwin's model differs slightly from Stigler's by emphasizing the distinction between ownership and control, thought important by industrial organization economists. In Baldwin's model the owners of the regulatory process are the political parties, but control resides in the hands of the regulators. Regulators want to maintain their organization and preserve their au-

tonomy and, to do so they must satisfy the conflicting parties in interest. This is done by arranging side payments. Organized interests are subsidized through the rate structure by unorganized interests. The regulation succeeds when it turns a zero sum game (rivalry before the legislature by competing interest groups) into a variable sum game. Baldwin, *Regulatory Agency and the Public Corporation*, pp. 5–16.

8. *See* William A. Jordan, "Producer Protection, Prior Market Structure and the Effects of Government Regulation," *Journal of Law and Economics*, 15 (April 1972), pp. 151–176.

9. *See* David Seidman, "The Politics and Economics of Pharmaceutical Regulation," in *Public Law and Public Policy*, John Gardiner, ed. (New York: Praeger, 1977) for an account of the FDA's sensitivity to Congressional pressures for stringent regulation. On the natural gas industry see Stephen Breyer and Paul MacAvoy, *Energy Regulation by the Federal Power Commission* (Washington, D.C.: Brookings, 1974). For an account of an air pollution control agency's efforts to organize proregulation pressure groups, see Paul Sabatier, "Social Movements and Regulatory Agencies: Toward a More Adequate and Less Pessimistic Theory of Clientele Capture," *Policy Science*, 6 (1975), pp. 301–342. Evidence casting doubt on the view that the Securities and Exchange Commission is a captured agency is presented in G. William Schwert, "Public Regulation of National Securities Exchanges: A Test of the Capture Hypothesis," *Bell Journal of Economics*, 8 (Spring 1977), pp. 128–150.

10. *See* Paul Joskow and Paul MacAvoy, "Regulation and the Financial Condition of Electric Power Companies in the 1970's," *American Economic Review Papers and Proceedings*, 65 (1975), pp. 295–301.

11. Joseph L. Bower, "Effective Public Management," *Harvard Business Review*, 55 (March–April 1977), pp. 131–140.

12. *Smyth v. Ames*, 169 U.S. 466 (1898).

13. Alan A. Altshuler, *The City Planning Process* (Ithaca, N.Y.: Cornell University Press, 1965), p. 77.

14. Sabatier, "Social Movements and Regulatory Agencies," p. 322.

15. For an analysis of attempts by consumer advocates to obtain influence and shape policy, see Mark V. Nadel, *The Politics of Consumer Protection* (Indianapolis: Bobbs-Merrill, 1971), pp. 155–218.

16. For a report of CAUSE's telephone campaign, see *Los Angeles Times*, August 8, 1976, Part IV, pp. 1–2. In a 1976 interview with the author of this book, Tim Brick, chairman of CAUSE, sought to differentiate the tactics employed by his group from traditional approaches:

"A large part of our function is to draw attention to hearings such as this. The *Times* and others accept only three strategies: entering lawsuits, lobbying the legislature and campaigning in elections. They think these are the only strategies that are justifiable and that's why they don't like us.

"We insist on using a variety of tactics for reaching people and affecting government. Publicizing initial hearings—making sure that lots of people and the media are there is one way. Creating outside pressure on the commissioners themselves is another. It is not as easy to create pressure on examiners except by packing the hearing room and physically showing concern.

"Choosing issues with potential for high visibility is half of it. Making sure they are highly visible is the other half. That's how we get participation and assure that the public interest will be achieved.

"For example, our real campaign began after the PUC [Public Utilities Commission] decision on the So Cal-Arco deal. [Note: the PUC authorized Southern California Gas to raise its rates as prepayment for Alaska natural gas to be supplied by Atlantic Richfield Company (Arco).] We urged customers to boycott bills, and we made a big papier-mâché shark [these events occurred at the time of the movie *Jaws*] that we named 'Rippy the Arco shark.' We wrote a play about how Arco was ripping off gas customers and put it on right under the nose of Arco [at their Los Angeles office]. The play and the shark generated a lot of attention from the media. We had debates on radio and TV. It was just like selling a book, and most of these talk shows are book sales programs. The TV stations, crews, and producers were sympathetic to our point of view. [Finally] Arco and So Cal Gas tore up their contract because they were getting so much bad press."

17. Mancur Olson, *The Logic of Collective Action* (Cambridge, Mass.: Harvard University Press, 1965).
18. Louis M. Kohlmeier, Jr., a reporter for *The Wall Street Journal*, expresses a typical view in his book *The Regulators*:

"[Every] . . . industry representative in Washington has a constitutional right to petition his government and none is lacking funds to petition effectively. It takes skill as well as cash. . . . No one need lose all the time, except the public. The politics of regulation is a merry-go-round propelled by the self-interests of industry that must ride for the preservation of revenues and profits; of regulators who want to keep their jobs; of members of Congress who need campaign cash; and of White House figures who want business confidence and use the regulatory agencies for political ends. The interrelationships of self-interests leave no room for considerations of the public interest. The agencies tend to become umpires not of consumer versus industry interests but of industry versus industry. Taken together, regulation tends to perpetuate the protection of industry and the disregard of consumer interests, including the consumer interest in industrial competition."

Louis M. Kohlmeier, Jr., *The Regulators* (New York: Harper & Row, 1969), pp. 80–81.

19. James Q. Wilson, "The Politics of Regulation," *Social Responsibility and the Business Predicament*, J. W. McKie, ed. (Washington, D.C.: Brookings, 1974), pp. 144–145.
20. I draw here on Sabatier, "Social Movements and Regulatory Agencies," pp. 319–320.
21. Gresham M. Sykes, *The Society of Captives* (Princeton: Princeton University Press, 1958), p. 42.
22. Wilson, "Politics of Regulation," p. 143.
23. Chester I. Barnard, *The Functions of the Executive* (Cambridge, Mass.: Harvard University Press, 1938), p. 73.
24. Alfred D. Chandler has utilized the concepts of "coordinating, appraising, and planning" in his masterful work *Strategy and Structure: Chapters in the History of Industrial Enterprise* (Cambridge, Mass.: MIT Press, 1962), pp. 19–51.
25. This proposition owes much to the work of Chester I. Barnard, particularly

Functions of the Executive, and James Q. Wilson's *Political Organizations* (New York: Basic Books, 1973). See especially Wilson, p. 9.

26. Wilson, *Political Organizations*, pp. 27, 31.
27. This view differs, of course, from that of some critics of pluralism. *See* Theodore J. Lowi, *The End of Liberalism* (New York: W. W. Norton, 1969).
28. For an elaboration of structural analysis within an industry that employs the concept of "strategic subgroup" see Michael E. Porter, *Competitive Strategy* (New York: The Free Press, 1980), pp. 126–155.
29. Allan P. Sindler, in his book *Huey Long's Louisiana* (Baltimore, Md.: The Johns Hopkins University Press, 1956), pp. 47–48, makes the following observation:

"Having secured statewide notoriety through his tilts with Standard Oil and the Governor, Huey next sought to develop a politically useful record on the New Public Service Commission, created by the 1921 Constitution in place of the Railway Commission. His golden opportunity came with the telephone rate controversy of 1920–22. As a result of hearings held in 1920–21, at which Long usually was absent, Taylor and John Michel, the other two members of the Commission, granted the Cumberland Telephone and Telegraph Company a 20 per cent increase in rates. In 1922, Huey Long, then chairman of the Commission, reopened the case. Ultimately, the rates were lowered, and since the reductions were made retroactive to the time of the company's application (1920), a large refund was assured to all telephone users. Huey Long became a state hero overnight. Long scored on other utility fronts as well, lowering the rates of the Southwestern Gas and Electric Company, of Shreveport streetcars, and of all intrastate railroads. Standard Oil of course, was not overlooked. Long helped persuade the 1922 legislature to abrogate the gentleman's agreement by enactment of a 3 per cent severance tax on petroleum obtained from Louisiana wells.

"By 1923, Huey Long thus was a personal force in state politics, an ambitious St. George who could not rest content merely with slaying utility dragons. At the age of thirty, the legal minimum for a gubernatorial candidate, Huey Long made his first bid for the highest elective state office."

30. For a discussion of entrepreneurial politics set in the context of the larger policy-making process see James Q. Wilson, *American Government: Institutions and Policies* (Lexington, Mass.: D. C. Heath, 1980), pp. 410–423.
31. Wilson, *Political Organizations*, p. 30; Barnard, *Functions of the Executive*, p. 217.
32. Barnard, *Functions of the Executive*, p. 219.
33. *See* ibid., pp. 114–115.

Chapter 2

ORIGIN OF
STATE UTILITY REGULATION

The most widely held view of the origin of state laws regulating public utilities is that these laws were passed over the strenuous objection of the utilities and that, once established, regulatory commissions had to contend with the sniping utilities that were forever striving to restrict the commissions' jurisdiction. This interpretation of events was expressed in one of the earliest textbooks on public utility regulation, published in 1933, and has been a part of the folklore of regulation ever since.[1]

In fact, state regulation received the public support of leading men in the electric power industry as early as 1898; after the movement for regulation was successful in establishing commissions in nearly every state, the electric utilities vigorously defended the jurisdiction of the commissions against encroachment by local and federal authorities. If other utilities saw state regulation as inimical to their interests, electric utilities in the leading states did not.[2] That the utilities sought to preserve their autonomy is uncontested; they did this not by opposing state regulation but by seeking it.

In 1907 there were essentially four possibilities for the future of the electric power industry: (1) control by the industry, with utility systems operated as private monopolies; (2) government ownership; (3) private ownership subject to state regulation; (4) private ownership subject to municipal regulation. Although the first possibility doubtless had the support of some businessmen, it was not a viable alternative. As electric utilities began to touch the lives of more and more people, there was nearly universal recognition that their development must be subject to some public influence. For different reasons, three disparate groups—the electric utilities, the National Civic Federation, and a number of reform governors—joined to support the concept of state regulation in place of either of the other two alternatives. Since the motivation for encouraging regulation differed among these groups as well as within them, it is important to examine the role of each.

The Utilities

It is impossible to adequately describe the contributions of the electric utilities to the movement for state regulation without discussing the role played by Samuel Insull, an immigrant to the United States from Great Britain who had served for a time as Thomas Edison's secretary and would later head Commonwealth Edison, Chicago's huge electric power system. In 1892, at the age of thirty-one, Insull arrived in Chicago to take control of Chicago Edison, which was then only one of the city's more than thirty electric companies.[3] Insull immediately set out to expand the business and fashion a monopoly of service in the city. He purchased and then retired from production the assets of competitors and obtained exclusive rights to buy the electrical equipment of every American manufacturer. In Chicago, the potential for growth was enormous: In 1892 only five thousand persons out of a population of one million used electric lights. With his success in raising capital, expanding capacity, and consolidating competitors, Insull soon became the city's most important electric utility executive. But his plans for further expansion were threatened five years after he arrived in Chicago by a group of extortionists on the city council known as the Gray Wolves.

Insull and the Gray Wolves

In 1897 the Gray Wolves had succeeded in thwarting certain plans of Charles Tyson Yerkes, Chicago's powerful traction magnate. Under state law, Chicago's public utilities were franchised by the city council for a maximum of 20 years. Because Yerkes's transportation system was a conglomerate of many smaller companies, he had to go to the city council for a renewal of his franchise every few years. The franchise arrangement made it difficult for Yerkes to sell the long-term bonds necessary for his system's further development and gave

the Gray Wolves repeated opportunities to harass Yerkes into paying for favorable votes. In 1896 Yerkes devised a bold scheme to rid
himself of the city council. He encouraged the legislature to establish a state regulatory commission which would take control of local
transportation companies away from city councils and, at the same
time, to extend the franchise to a period of 50 years. Yerkes apparently made $500,000 available to secure the votes of members of the
1897 legislature; but his bribery attempt was exposed, and his plan
to remove utilities from local control was ultimately defeated by an
odd coalition of crooks and reformers.[4]

Yerkes's bill to extend the franchise period to 50 years passed, but
city councils retained the power to grant franchises. Within a year
even this law was repealed, but while it was on the books the Gray
Wolves had an opportunity to focus their special skills on Samuel
Insull. A few days after the franchise bill had passed, the Wolves
sent emissaries to Insull to inform him of their plan to franchise a
new company, the Commonwealth Electric Company, which they
would own. The Wolves intended the company to be nothing more
than a dummy corporation; their sole purpose was to impel Insull to make a considerable offer to buy the company's franchise.
Insull flatly refused their first offer to sell, so the Wolves took steps
to turn the dummy corporation into an active competitor. It was
then that they learned Insull held exclusive rights to the purchase of
all American electrical equipment. Without equipment Commonwealth Electric could scarcely compete. Four months later, the
Wolves agreed to sell the franchise to Insull for $50,000, a fraction of
what they had hoped to get out of him.[5]

Insull Advocates State Utility Regulation

Insull's experience with the Gray Wolves must certainly have affected his attitude toward local control of utilities, for the following
year (1898), as president of the National Electric Light Association
(NELA), the electric industry's trade association, he advocated the
elimination of competitive franchises and the establishment of a
system of legislative controls of rates and service.[6] Competition, he
argued, had not lowered the price of electricity but had only made
investments riskier and costs higher. He further asserted that to
acquire capital at low interest rates utilities needed to be protected

from competition, but that in return for exclusive franchises they must be willing to accept public control. Alluding to the European regulatory experience, Insull said, "The more certain this protection is made, the lower the rate of interest and the lower the total cost of operation will be, and consequently, the lower the price of the service to public users. If the conditions of our particular branch of public service are studied in places where there is a definite control, whether by commission or otherwise, it will be found that the industry is in an extremely healthy condition, and that users and taxpayers are correspondingly well served."[7]

Despite his standing in the industry, Insull's position was still too advanced for the NELA to endorse. He did manage to establish a committee on legislative policy and appointed himself, Edgar H. Davis of Williamsport, Pennsylvania, H. M. Atkinson of Atlanta, Samuel Scovil of Cleveland, and Charles R. Huntley of Buffalo to study the question. However, the committee had little impact on the NELA's policy during the next six years. One of Insull's fellow committee members, Edgar Davis, later expressed hostility toward the idea of government regulation in his presidential address before the NELA convention in 1905. "The one great and constant menace to the industry is unwise, burdensome and restrictive legislation by the municipality and the state. . . . the power to regulate contains the germ of the danger of confiscation, in whole or in part."[8]

Insull's initial arguments on behalf of state regulation—that it was preferable to local competition with its associated political bargaining for municipal franchises and that it would serve as a means to reduce risk in the industry and lower its cost of capital—failed to persuade large numbers of his fellow executives. But another threat to the industry—municipal ownership—did. According to Insull's biographer, Insull had no real fear of municipal ownership; indeed, he advocated a government-owned system for England and even spoke occasionally for municipal ownership in the United States.[9] Other electric utility executives were not so sanguine about municipal ownership, however; they were terrified that it would wrest from them control of their businesses. Insull saw in this fear the basis for a new campaign for state regulation. With others, in 1904 he formed the Committee on Municipal Ownership under the auspices of the NELA with the intention of presenting regulation as the only alternative to government ownership. Two years later the committee was renamed the Committee on Public Policy.

Report of the NELA Committee on Public Policy

The report of the Committee on Public Policy, read at the Washington, D.C., convention of the NELA in June of 1907, is the most important statement on the relationship of government and public utilities ever issued by the electric power industry, for it has since become the basis of the industry's position on government regulation. It was particularly timely because in that same year the first modern public utility commissions were established in New York and Wisconsin. The report's acceptance by the convention was due in large to the prestige of its authors. The nine-member Committee on Public Policy was comprised of, among others, three past presidents of the NELA, Henry L. Doherty of Denver, Samuel Insull, and Charles E. Edgar of Boston; two future presidents, J. W. Lieb, Jr., of New York, and Joseph B. McCall of Philadelphia; and Samuel Scovil and Alex Dow, the head of Detroit Edison. The prestige of the committee was attested to by the convention's presiding officer, who remarked, not altogether gratuitously, "It has been said that our public policy committee is the strongest committee that has ever been formed on behalf of any association, and I do not think anyone will be inclined to differ with that opinion."[10]

The members of the Committee on Public Policy were under no illusion that the electric power industry could escape the demand for public scrutiny of operations and rates. Their report stated, in part:[11]

> *That which is now uppermost in the public mind is public supervision and control. In the judgment of your committee some form of such supervision and control is inevitable in many if not all of the important states of the Union, and we believe it should be welcomed by the parties in interest, provided it is put, as we believe it can be, in such form as to preserve the rights and properties of the companies as well as to promote the interests of the public. The practical question is not whether there is to be such regulation and control as it is what the nature and form of them are to be.*

In the view of the committee, if municipal ownership were to be avoided, the form of control needed to be public regulation. In introducing reports on municipal ownership and public regulation and control the committee wrote:[12]

> *The [two] subjects . . . are intimately connected with each other. Neither can be adequately discussed without reference to the other. Indeed, one is the alternative of the other. Municipal ownership is demanded largely because of the absence of proper regulation and control. Public regulation and control, if efficient, removes the neces-*

sity or excuse for municipal ownership by securing fair treatment for the public.

The subcommittee report on municipal ownership, written by Alex Dow (head of Detroit Edison) and Samuel Insull, developed the theme of public regulation, noting that "the propaganda in favor of [municipal] ownership is losing its vitality . . . to a large extent [because of] the rapidly approaching culmination of the idea of public regulation." The public, Dow and Insull argued, is not as interested in the form of ownership or control as it is in good service at reasonable rates. If public regulation could offer the public fair treatment, they asserted, the public would accept it as an alternative to government ownership. But in a warning that rang of prophecy, Dow and Insull stressed:[13]

Municipal ownership is not discredited; it is merely forgotten. It would be a serious error to assume that the present movement of public sentiment toward public regulation signifies that municipal ownership is now or is soon going to be consigned to the limbo of discredited theories, along with such crazes as free coinage of silver at a ratio 16 to 1. If public regulation shall fail to establish a good understanding between the corporations operating public utilities and the customers of those corporations, we shall inevitably have a revival of the cry for municipal ownership.

Having established municipal ownership as the only alternative to public regulation and having presented the case for the latter, the full committee sounded the following cautionary note:[14]

While agreement may be reached upon the general principle that public regulation and control of public service corporations is desirable in the public interest, and is not necessarily inimical to the safety and value of corporate investments, it is another and much more difficult matter to agree upon the nature and scope of it.

A review of legislation pending in New York and other states to create utility commissions prompted the committee to express admiration for provisions requiring staggered terms of office for commissioners for the sake of continuity, but it expressed concern over the breadth of discretion granted to regulatory commissions. Much would depend, argued the committee, on the quality of the persons appointed to commissions; it noted that undesirable commissioners could inflict "tremendous harm" on utility companies. The committee stated: "Indeed, it is difficult to suggest any other political machine which would be anything like as effective in its operations and as baneful in its consequences."[15] Every effort should be made to obtain capable and honest persons as commissioners, the commit-tee urged: "To that end we would advocate long terms and secure tenure of office, with adequate salaries, sufficient as far as possible to remove the element of self-sacrifice in the acceptance and incum-bency of the offices."[16] Interestingly, the committee went on record as supporting the proposal that regulated companies be assessed for the salaries of the commissioners.

The report of the Committee on Public Policy concluded that the administrative discretion provided by the New York law might be necessary for the approval of the concept of public regulation, but the committee stated it indulged the hope that "even if the first legislation upon this subject is destined to take the form of nonautomatic or discretionary regulation and control, the results of its workings in practice may be the evolution of a highly developed and practicable system of automatic or semi-automatic regulation."[17]

Industry Reactions to Utility Regulation

The reports of the Committee on Public Policy and its subcommit-tees on regulation and municipal control provide a great deal of information about early industry attitudes on the role of government in the utility business. Their authors were as certain that the move-ment for government control was unstoppable as they were that the movement was linked to larger forces, including the growing de-mand for more stringent control of railroads.[18] The authors were equally convinced that they needed to support the movement for state commissions to protect their businesses. "The wise course would seem to consist not in an attempt to stem the tide of public opinion but rather in seeking to evolve and define that method of public supervision and measure of control which will permit . . . the companies [to] conduct their business in the most progressive and enterprising manner," wrote Samuel Scovil and Joseph B. McCall.

The NELA Subcommittee on Public Regulation and Control summarized its position in the following three conclusions:[19]

First: *That the National Electric Light Association should favor properly constituted general supervision and regulation of the electric light industry.*

Second: *That if state commissions be constituted, they should be appointed in that manner which will give them the greatest freedom from local and political influences, to the end that their rulings shall be without bias.*

> Third: *That state commissions should be clothed with ample powers to control the granting of franchises, to protect users of service against unreasonable charges or improper discriminations, to enforce a uniform system of accounting, and to provide for publicity. If the state provides for publicity on the one hand, on the other hand it should safeguard investments. Regulation and publicity would be a grievous wrong unless accompanied by protection.*
>
> *In the conferring of these powers, great care should be taken. Regulation in and of itself necessarily introduces a factor of resistance to the adoption of new methods and to progress generally. Hence in the interest of lower costs, and consequently lower charges to users of service, the functions of the commissions should be confined within strictly regulative lines.*

Although the subcommittee reports were not voted on separately, they received the approval of the full committee, and when a resolution calling for the acceptance and adoption of the report of the Committee on Public Policy was placed before the association, the report received a unanimous vote of approval.[20]

Even before the 1907 NELA convention, a number of electric utility executives were at work in their states supporting the concept of regulation. The executives' support was neither uniformly enthusiastic nor uniformly effective, but the generally accepted notion that the most important leaders of public utilities were forced into a system of state regulation is simply wrong. In Wisconsin, for example, Henry C. Payne, the vitriolic, anti-progressive state Republican boss who also happened to be the vice president of the Milwaukee Electric Railway and Light Company, actively lobbied for a state regulatory commission.[21]

After the 1907 convention the electric power industry's executives were even more committed to the concept of state regulation than they had been before, and as they returned to their home states they carried the message of state regulation with them. This is not to say that all industry executives supported regulation or viewed regulation as the only alternative to public ownership; they did not. At the very next NELA national convention the association's 1908 president, Dudley Farrand, expressed concern that the movement for regulation might go too far and unnecessarily restrict the industry's autonomy:[22]

> *The interests that we represent have been subjected to an agitation having an indefinable source but extending practically all over the country. This agitation, wearing the cloak of reform, has for its objective the creation and establishment of a multiplicity of commissions,*

*national, state and municipal, and seeks not only a reasonable regula-
tion and control, as stated in some of the arguments in favor of the
movement, but it actually prevents progress by unreasonable interference with private interests, stifles individual initiative, and in many
instances favors acts that are no less than actual confiscation.*

Farrand counseled his fellow executives to "feel free to protest . . .
and to defeat the enactment of measures designed for political purposes and tending to undermine the very foundations of our investments."[23] There were, of course, utility managers who agreed with
and followed the strategy suggested by Farrand. In New Jersey, for
example, the general counsel for the public service corporation of
New Jersey appeared before the state senate in March 1908 to protest the formation of a public utility commission modeled on that of
New York. He claimed the New York commission had precipitated
the bankruptcy of the New York City street railroad system within
months after the law creating the commission had been passed.[24]

In 1908 the New Jersey Public Service Corporation owned nearly
all of the street railways and gas and electric companies in the state,
so it is not surprising that with so many interests some of its managers opposed regulation. Not all of the corporation's executives
shared this view, however, as indicated by a statement supporting
regulation by Walton Clark, a member of the corporation's board of
directors and vice president and general manager of the United Gas
Improvement Company of Philadelphia, a concern that controlled
the gas works of about fifty United States cities and towns. In a
statement released in New York City in June 1907, Clark reported
that he favored state regulation as an alternative to municipal
ownership, "Municipal ownership has not proven equal to private
ownership in benefits to the consumer, citizen, or city . . . I recommend state regulation and protection of public service companies,
provided by statute, and as far as possible automatic in its application
and operation."[25]

A number of leading electric utility executives shared Clark's view
and pressed for the establishment of state commissions in the years
following the 1907 NELA convention. In December 1907, Alex Dow
appeared before a committee of the Michigan constitutional convention on behalf of Detroit Edison to support the establishment of a
state regulatory commission, arguing that such a body could provide
an effective "substitute for competition."[26] Some of the Detroit
newspapers differed with Dow, however, seeing the proposal as a

strategy by which the utilities could avoid municipal control and frustrate the movement for municipal ownership—which of course it was. In 1907 the lines of support and opposition in Michigan were so indefinite that it was difficult to make out who belonged to the two coalitions that were the main actors in the dispute over municipal control. A state commission to regulate electric rates was finally blocked by a group of home rule advocates and by those who favored extended state utility control and municipal ownership, with help from the railroads and representatives of other utilities such as the Detroit City Gas Company, which opposed state regulation.[27] Dow said that his proposal for state regulation was threatened with defeat "because it was found in bad company."[28] Instead of state regulation, the 1908 Michigan constitution contained a home rule provision that allowed cities to set rates and to grant franchises to electric power companies and other utilities. Michigan home rule law was amended in 1909 to give some jurisdiction for electric utilities to the state railroad commission. The new law provided that the commission "shall in no case have power to change or alter the price for electricity fixed or regulated by or under any franchise . . . granted by any city, village or township." In other states that adopted similar home rule provisions, commissions did not receive the authority to alter rates until after the First World War.[29]

In California, where the effort to regulate electric utilities was greatly influenced by public discontent over state regulation of the railroads, the largest utilities were in "the vanguard of those clamoring for its passage."[30] At the 1909 convention of the Pacific Coast Gas Association, the association's president urged its member utilities to work for the adoption of state regulation "on account of the growing desire on the part of our city governments to regulate our affairs."[31] A legislative committee composed of representatives of the Pacific Gas and Electric Company, the Los Angeles Gas and Electric Company, and the Pacific Lighting Company was appointed by the same convention to work on behalf of state regulation. The next year the association's president came out even more strongly in favor of state regulation. "Local men are not capable," he said, "of fixing rates for public service corporations without prejudice." He then asserted that state regulation would provide "a much more settled condition as to competition."[32]

When the bill to enlarge the California railroad commission's jurisdiction to cover all utilities was before the state legislature in

1911, the affected utilities sent their representatives to speak in support of its passage. John Britton of Pacific Gas and Electric Company and Tiery Ford of the Sierra and San Francisco Light and Power Company told the *San Francisco Examiner*, "We are glad to be regulated for our own sake."[33] Had the utilities opposed the public utilities act it would not have passed both houses of the legislature as easily as it did. The month the law went into effect, a high official of the Southern California Edison Company voiced his conviction that the new commission would provide "absolute stability of our securities and protection from unnecessary competition."[34]

Electric utilities in every state were not unanimously in support of state regulation, but neither were they wholeheartedly in opposition, as the conventional view of history would have it. In a number of the largest states, utilities vigorously supported the state regulation concept, and their leaders cautioned against fighting regulation. Nowhere is this caution more forcefully stated than in a 1910 speech by Samuel Insull given before a convention of H. M. Byllesby and Company, an electric utility management and engineering consulting firm. The convention had earlier heard Charles G. Dawes, president of the Central Trust Company of Illinois, attack the idea of state regulation and urge utility executives to fight its adoption. Insull's response was as follows:[35]

> *Our friend, Mr. Dawes, has referred to the tendency of the times so far as legislation is concerned. While as an abstract proposition I think it is very laudable for us to cheer the idea that we should go out and fight any curtailment of our liberty of action, as suggested by Mr. Dawes, yet, as a practical, everyday proposition, and as a necessity, we have to face the views of the various communities of the states in which we are engaged.* We should bear in mind, above everything else in the operation of our business, that we cannot afford to place ourselves in opposition to public opinion. [*emphasis added.*] *If we are to maintain values of the securities for which we are responsible, and to increase those values, we should rather bend our energies to find some means of operating our business to meet the conditions that will undoubtedly confront us in most of the states . . .*
>
> *I think it was some twelve years ago that I first tried to voice the idea that our business is a natural monopoly and that we must accept, with that advantage, the obligation which naturally follows, namely, regulation.*
>
> *For my own part, I cannot see how we can expect to obtain from the communities in which we operate, or from the state having control over those communities, certain privileges so far as a monopoly is concerned and at the same time contend against regulation. Further, I*

think that regulation of the price of our product must be followed by regulation as to the issuance of securities, because our price must depend on the fixed charges we have to pay; and I cannot see how those fixed charges can be kept down within proper limits unless the authorities, in some way, either the community or the state, have the right to state the terms on which these securities shall be issued.

I am not proposing to get into a controversy with Mr. Dawes on this subject, but I think we will greatly strengthen our position, and greatly strengthen the securities issued against our business, if we accept the inevitable, and instead of trying to oppose the handwriting on the wall, try, rather, to direct the tendency so indicated toward getting legislation which will enable us to conduct our business in a way satisfactory to the public.

The National Civic Federation

The success of the movement for state regulatory commissions is partially attributable to the managers of leading electric power companies and other utilities, but success was by no means gained by their efforts alone. Also important in the state commission movement was a group of civic reformers, some of whom belonged to the National Civic Federation.[36] If utilities sought regulation to protect themselves from competition and municipal ownership, the National Civic Federation and other participants in the coalition for state commissions had different motivations and correspondingly different views of the proper role of government regulation.

Noting that no impartial or scientific study had ever addressed the relative merits of private and public ownership and operation of public utilities, the National Civic Federation announced its intention in September 1905 of conducting such an investigation on both sides of the Atlantic. A commission of one hundred and fifty widely known corporation heads, labor leaders, and publicists was formed to raise money for the study; its executive committee included Samuel Insull; John Mitchell, president of the United Mine Workers; and Louis Brandeis, at that time practicing law in Boston. A separate "committee on investigation" consisting of 21 members was formed and charged with the duty of carrying out the study. To insure "the greatest possible degree of impartiality," the committee was comprised in equal numbers of persons who had expressed opinions in favor of municipal ownership or against municipal ownership and persons who had expressed no opinion.[37] A friend of

Insull's, Charles L. Edgar, president of Boston Edison, was a member of the committee; so too were Frank Parsons, president of the National Public Ownership League, and John R. Commons, professor of economics at the University of Wisconsin.

The Report of the NCF Committee on Investigation

In the summer of 1907, after two years of study, the committee on investigation released its findings, with 19 of 21 members agreeing to its majority report. That so diverse a group was able to issue a consensus report at all is remarkable. The following principle had the unanimous support of the committee:[38]

> *Public utilities are so constituted that it is impossible for them to be regulated by competition. Therefore, they must be controlled and regulated by the government; or they must be left to do as they please; or they must be operated by the public. There is no other course. None of us is in favor of leaving them to their own will, and the question is whether it is better to regulate or to operate.*

As for the question of private or public ownership of utilities, the committee's majority concluded that it could not take a general position. But it did agree on two points:[39]

> *(1) Public utilities, whether in public or in private hands, are best conducted under a system of legalized and regulated monopoly;*
> *(2) Private companies operating public utilities should be subject to public regulation and examination under a system of uniform records and accounts and of full publicity.*

Perhaps even more interesting than the two points made by the committee are the reasons the committee gave for being "unable to recommend municipal ownership as a political panacea." Their chief concern appears to have been that municipal ownership would perpetuate the power of urban political machines by increasing the number of city jobs available for patronage distribution. The committee did not find political machines operating in British local governments.[40] Instead in England and Scotland they "found a high type of municipal government, which is the result of many years of struggle and improvement. Businessmen seem to take a pride in serving as city councillors or aldermen, and the government of such cities as Glasgow, Manchester, Birmingham, and others includes many of the best citizens of the city." Such conditions were "distinctly favorable to municipal operation."[41]

By contrast, many American cities did not enjoy such favorable conditions, in large part because of the corrupting influence of public service corporations that pandered to politicians. "There seems to be an idea with many people," the committee wrote, "that the mere taking by the city of all its public utilities for municipal operation will at once result in ideal municipal government. . . . We do not believe that this of itself will accomplish municipal reform."[42] According to the committee, successful municipal operation of public utilities "depends upon the existence in the city of a high capacity for municipal government" and until that capacity develops it is better to reassert control over the private operation of utilities through regulation. The committee stated: "With the regulations we have advised, with the publication of accounts and records and systematic control, the danger of the corruption of public officials is very much reduced."[43]

Public service corporations were as much the victim as the source of municipal corruption, in the view of Walter L. Fisher, author of the committee's summary report on the state of the American city and one-time president of the Municipal Voter's League of Chicago. In that fact, according to Fisher, lay the basis for the utilities' agreement with the principle of regulation:[44]

> *It is . . . claimed and . . . sometimes true that the political activity of the public service corporation, after it has been granted . . . a franchise . . . is due to necessity for self-protection against unreasonable attacks by public officials inspired by corrupt or demagogic motives. To remove such a justification where it exists and to render public regulation intelligent and fair as well as effective, is one of the results hoped for from the creation of . . . expert commissions under legislation such as . . . the public utility statute in New York.*

But the committee went beyond recognizing the industry's position in the following two recommendations, aimed at enabling cities to obtain and operate utilities if they should desire to do so:[45]

> *(1) Franchise grants to private corporations should be terminable after a fixed period and meanwhile subject to purchase at a fair value;*
> *(2) Municipalities should have power to enter the field of municipal ownership upon popular vote under reasonable regulation.*

The purpose of the provisions was not so much to foster public ownership as to control private companies, as is made clear by Walter Fisher:[46]

> *That the city should have legal power to own and operate its utilities is . . . generally conceded and enabling legislation has recently been enacted in many states. This has been due not wholly to a demand for actual public ownership, but in large part to the recognition of the fact that public regulation can never be really effective unless there exists the present power of the city to take over the particular utility upon proper terms if its private owners fail to operate and develop it . . . as an effective agency for the public service.*

The NCF Report and State Legislation

The report of the National Civic Federation proved to be an impor- tant stimulus for state regulation, not only because of the wide pub- licity the investigation and its findings received but also because its findings served as the outline for state legislation even before they were published. In Wisconsin, John R. Commons was asked by former Governor Robert M. La Follette and Herman L. Ekern, Speaker of the assembly, to draft a law extending state regulation to municipal and interurban public utilities. La Follette knew that Commons was at work on the National Civic Federation's investiga- tion and anticipated that his recommendations would closely resem- ble the Federation's. They did, and the draft bill Commons pro- duced became the Wisconsin law that later served as a model for the laws of many other states. Of the influence of the National Civic Federation, Commons writes, "It was in the midst of winding up [the] Civic Federation's report that I worked during six months on the public utility law. . . . I adopted nearly the whole of the recom- mendations signed by nineteen of the twenty-one members of the investigating committee of the Civic Federation. I did not, on my own initiative, introduce anything new in drafting the bill."[47]

The involvement of the National Civic Federation in the devel- opment of state regulatory commissions did not end in 1907. Four years later the NCF commissioned another study to produce a model bill for the "Public Regulation of Interstate and Municipal Utilities." The draft bill, embodying in greater detail the recom- mendations of the 1907 report, was completed by 1913 and was widely circulated among state legislatures considering public utility legislation that year.[48]

It is always difficult to assess the importance of the contribution of one organization to the success of a piece of legislation as complex as the laws that created state utility commissions, but Shelby B.

Schurtz, a contemporary critic of the utilities and of state regulation, was moved to say, "The most potent influence in the passage of all the recent public utility legislation has been the National Civic Federation."[49]

The Reform Governors

In addition to the utilities and the National Civic Federation, the movement for state regulation of electrical utilities had the support of leading progressive governors. Indeed, a number of governors used regulatory issues as a catapult to national prominence. Four governors in particular, Charles Evans Hughes of New York, Robert M. La Follette of Wisconsin, Hiram Johnson of California, and Woodrow Wilson of New Jersey, established reputations that made each a serious contender for the presidency on the issue of government control of private enterprise. Each was the governor of his respective state when the key battles to create state commissions to regulate utilities were fought.[50] Although the four governors differed in many ways and eventually opposed one another for the nation's highest office, they had similar views on the regulation of public utilities when the movement for state commissions was in its infancy.

Charles Evans Hughes and the New York City Utility Rate Investigation of 1905

The early political career of Charles Evans Hughes merits somewhat closer attention than the careers of the others. It was Hughes who sponsored the legislation in New York which created the first modern state utility commissions, and his political success provided a conspicuous example for scores of political entrepreneurs anxious to create favorable public opinion to launch careers of their own.

Oddly enough, Hughes's political style was characterized by nothing so much as his distaste for politics and its practitioners.[51] As governor of New York, Hughes refused to bargain with the state legislature or members of his party on key appointments and his legislative program, preferring to appeal directly to the public for the support he needed. Hughes was not a power broker but rather something of what might be called today a media "event." From 1905 until he was appointed to the United States Supreme Court in

1910, Hughes provided the New York press with its best in-state copy. In his skillful use of the print medium he anticipated the political use of radio by Franklin D. Roosevelt and the mastery of television by John F. Kennedy.[52]

Hughes's discovery of the force of public opinion and of his ability to shape it came not so much by design as by chance. He was a prominent New York City lawyer and former Cornell and Columbia law professor when, at the age of forty-three, he was asked to take charge of the first of two great investigations into corporate affairs which were to make him a national figure.

At the turn of the century New York City was in the midst of a battle over utility rates and service. In 1902 Mayor Seth Low rejected all bids on a public lighting project, charging that the companies were gouging the people and that they had an obligation to provide adequate service at a reasonable price. The utilities disputed this, claiming that the seller alone had the right to set prices.[53] Low was defeated in his bid for reelection in 1903, but his successor, George B. McClellan, continued to tilt against the utilities and to talk about setting up a municipal lighting plant. Meanwhile, the City Club and the Merchants Association began agitating for an investigation into the rates of gas and electricity. Tammany Hall stood solidly against an investigation, but mounting pressure from the city's major daily newspapers finally succeeded in prompting the state legislature to authorize an inquiry. The legislative investigation committee was to be chaired by Senator Frederick C. Stevens, a millionaire who had once owned an electric lighting company in Washington, D.C., but who was committed to conducting an impartial inquiry. Stevens was determined to attract the services of the best New York lawyer he could find to serve as chief counsel to the committee. Wherever he asked, Stevens was told that Charles Evans Hughes was the man he was looking for.[54]

Hughes at first declined the offer to be the committee's chief counsel. "I knew nothing of the gas business," he later wrote. "I had little or no confidence in the integrity of legislative investigations, and I feared that, with the great financial interests involved, the investigation would be thwarted in some way, and I should be in a position of apparent responsibility and debited with a conspicuous failure in a matter in which there was intense public interest."[55] Nor did Hughes enjoy the thought of the exposure he was certain to get from the press. In his autobiographical notes he maintained that he "hated the idea of work where the public eye would be upon every

step, with the newspapers keen on the scent for any political in-
triguing and the Hearst press ever ready to make sensational
charges."[56]

Not to be denied, Stevens pressed Hughes to accept the appointment, and eventually, after Hughes was convinced of the purity of
Stevens's motives and after he had won a promise of complete independence, Hughes consented to head the inquiry. His selection was
announced to the press on March 24, 1905. Before the end of the
month, the investigation commenced and witnesses were called.

The press remained skeptical. Hughes belonged to the same
church as John D. Rockefeller, who supposedly controlled the gas
monopoly. (He had even taught Rockefeller's son in his Sunday
school class.) Furthermore, he was a friend and former law partner
of the attorney retained by the utilities to represent their interests in
the investigation. Few journalists thought the investigation would
be anything other than a "whitewash expedition," as so many earlier
legislative investigations had been.[57]

For most people, the spate of revelations produced by Hughes's
methodical examination came as a complete surprise. Hughes's
public image changed overnight. Whereas before the press had cas-
tigated him as a "trust lawyer," they celebrated him as a major new
public figure. The *New York Evening Mail* described Hughes as "a
large man, not burly, but with the appearance of one who is built on
big, strong lines. He looks strong. His shoulders are square, his
limbs solid, his teeth big and white and his whiskers thick and
somewhat aggressive."[58] Newspapers made much of Hughes's
whiskers. Cartoonists depicted the whiskers as the straws of a
broom, sweeping corruption out and good government in.[59] The
New York World confided to its readers that "in real life [the whis-
kers] are broader, braver, bigger, bushier" than they appeared in
photographs and that "when in action they flare and wave about
triumphantly like the battle flag of a pirate chief."[60]

Among other things, Hughes succeeded in proving (1) that Consolidated Gas Company had charged New York City $80,000 for the
same amount of electric current for which it had charged private
customers $25,000; (2) that electricity sold to the city for 4.86 cents
per kilowatt-hour was earning a margin of 2.44 cents per kilowatt-
hour; (3) that New York Gas and Electric Light Company sold current to residential customers at an average rate of 8.042 cents per

kilowatt-hour although the current was produced at a cost of 3.664 cents per kilowatt-hour and that some customers paid as much as 15.00 cents per kilowatt-hour for their electricity; and (4) that Consolidated Gas had sought to keep rates up and taxes down by reporting its taxable property at only 75 per cent of the valuation it used for rate-making purposes.[61]

Hughes completed the investigation in three weeks and, with remarkable energy, produced his report a week later, in time to take it and his proposals to Albany before the close of the legislative session. Hughes made two recommendations for the price of electricity and gas which won immediate applause from the public: He proposed that the maximum price of electricity be dropped by one-third, from 15 cents to 10 cents, and that the price of a thousand cubic feet of gas be dropped by one-quarter from one dollar to 75 cents.

More importantly, Hughes rejected the concept of controlling public service corporations by franchise provisions and general laws and called instead for a public service commission to supervise the activities of gas and electric utilities in the state:[62]

> *The gross abuse of legal privilege in overcapitalization and in the manipulation of securities for the purpose of unifying control and eliminating all possible competition shows clearly that there can be no effective remedy by general legislation or through ordinary legal proceedings, and that for the protection of the public there should be created a commission with inquisitorial authority, competent to make summary investigations of complaints, to supervise issues of securities and investment in the stocks or bonds of other companies, to regulate rates and to secure adequate inspection, or otherwise enforce the provisions of the law.*

Although the 1905 legislature responded to Hughes's findings by creating a state commission of gas and electricity, the commission's rate-setting powers were few and were quickly constrained by a court challenge that blocked the recommended reduction in the price of gas.[63]

The gas and electric inquiry brought Hughes to the attention of New York State within a few weeks; scarcely four months passed before another investigation—this time into the insurance industry—made Hughes a national sensation almost as fast. Hughes was well aware of the potential an inquiry into the insurance industry's practices held for its principal investigator. When he learned of his appointment to the inquiry, he was in the Alps, where he had gone with his family to relax after the gas and electric investigation. "It would be the most tremendous job in the United States," he enthused to his wife.[64]

And it was. New York life insurance companies carried policies on millions of people throughout the country. Under Hughes's leadership, discoveries of instances of ineptitude, political intrigue, and fraud were more numerous even than those made in the gas and electric industries. So threatening was the Hughes investigation to the insurance industry and its political allies that midway through the proceedings they succeeded in securing for Hughes the Republican nomination for mayor in the hope that it would divert the investigation. Hughes knew nothing of the nomination until confronted by an official notifying committee one evening at his doorstep; he flatly refused to accept it.

Hughes and State Utility Regulation

In 1906, with the insurance investigation completed and the probability of a gubernatorial campaign by William Randolph Hearst increasing, Hughes was more willing to accept the nomination for governor than he had been to accept the nomination for mayor. He saw his nomination as a call to duty to protect the state against a potentially disastrous Hearst regime. With others, Hughes believed that Hearst's yellow journalism had contributed to the frenzy that had led to President McKinley's assassination. After Hearst had received the Democratic nomination, Hughes declared, "The man who would corrupt public opinion is the most dangerous enemy of the state. . . . As against reckless denunciation, I set fair criticism. As against indiscriminate attacks upon business organizations, I set a serious and determined attempt to ascertain the evil and to remedy it."[65]

Hughes, of course, defeated Hearst. In his inaugural address, he reemphasized his commitment to fair and evenhanded government, "I believe in the sincerity and good sense of the people. I believe that they are intent on having government which recognizes no favored interests and which is not conducted in any part for selfish ends."[66]

The day after his inauguration Hughes submitted to the state legislature his legislative program, the most important component of which was a proposal to abolish the state's board of railroad commissioners and commission on gas and electricity and New York City's rapid transit board, and to create in their places two comprehensive state public service commissions, one for New York City and one for the rest of the state. The governor was to appoint the members of both commissions and was to have the power to remove any commissioner at will. Hughes felt deeply that unless the governor retained strict control over the appointments and the removal process, the commission that regulated utilities in New York City, especially, would come under the influence of the machine and the anti-progressive politicians he had struggled against.

The New York legislature was not as eager to curb the machine's power as Hughes was, however, and challenged the worth of Hughes's proposal. But they were no match for the Governor. Hughes carried his campaign for the public service commissions bill to gatherings of citizens throughout the state and used every possible occasion to win support for the bill. To an audience of businessmen he ridiculed the charge that the regulatory movement would disrupt industry:[67]

Will anyone suggest to an intelligent audience that American citizens are in revolt against their own prosperity? What they revolt against is dishonest finance. What they are in rebellion against is favoritism which gives a chance to one man to move his goods and not to another; which gives to one man one set of terms and another set to his rival; which makes one man rich and drives another man into bankruptcy or into combinations with his more successful competitor. It is a revolt against all the influences which have grown out of an unlicensed freedom, and of a failure to recognize that these great privileges, so necessary for public welfare, have been created by the public for the public benefit and not primarily for private advantage.

When a prominent state legislator attacked the public service commissions before his hometown chamber of commerce and claimed that he spoke as a citizen "under no retainer from the railroads," Hughes, who was sharing the dais, rose to exclaim:[68]

In distinction from my learned friend, I am here under a retainer. I am here retained by the people of the State of New York, to see that justice is done, and with no disposition to injure any investment, but with every desire to give the fullest opportunity to enterprise, and with every purpose to shield and protect every just property interest.

I stand for the people of the State of New York against extortion, against favoritism, against financial scandal, and against everything that goes to corrupt our politics by interference with the freedom of

*our Legislature and administration. I stand for honest government
and effective regulation by the State of public-service corporations.*

Hughes's personal appearance, his rhetoric, and his legislative program received such favorable press attention that the Legislature,
cowed, gave him the commissions without compromise on May 22,
1907.

Robert M. La Follette and "Scientific" Regulation

A little more than a month later the Wisconsin legislature followed
New York's lead by introducing its own bill authorizing the state
railroad commission to regulate the financial activities, establish service standards, and fix the rates of heat, water, light and power, and
telephone companies. The real fight for a state regulatory commission had taken place a little more than two years earlier in the
legislative election of 1904 when Governor Robert M. La Follette
had broken with tradition and stumped the state, vigorously campaigning against legislators who were supportive of the railroads. His
efforts were successful: In 1905 the Wisconsin Legislature voted to
create a commission to control the railroads. As John R. Commons
wrote, "What remained for the Legislature of 1907 was simply to
bring all other public utilities under the same commission and to
deal with the questions of franchises and the relation of municipal
governments to the State commission."[69]

For La Follette, the question of the proper jurisdiction—state or
local—for control of utilities was an important one, and he was
determined to have the question resolved in favor of the state. He
had two reasons for this stance: (1) he thought the state could
provide more expert and therefore more careful supervision of industry; and (2) he viewed local governments as hopelessly corrupt.
In this view he was supported by electric utilities which, as has been
pointed out, unlike the railroad welcomed state regulation as a protection against local political demands and the threat of municipal
ownership.

Unlike Governor Hughes, Governor La Follette endeavored to
replace one political machine with another for the purpose of
"emancipating the party from the domination of the established 'system' and . . . [making it] more directly responsive to the popular
will."[70] To that end La Follette cultivated personal relationships
with people throughout the state and solidified his organization

through the use of patronage. He is reported to have maintained a card file of all the men in Wisconsin he had ever met, arranged by locality so that he could look them up and refresh his memory when he planned to visit their communities.[71]

These methods were successful in securing La Follette's election as governor, but although he had a larger personal acquaintance than any other man in Wisconsin, his support came mostly from rural machines. The local urban machines used their power to elicit "contributions" from the public utilities and used the money to perpetuate their control. By removing public utilities from the control of local machines and creating a state commission, La Follette sought both to weaken his opposition and to provide for expert, scientific, and therefore, by his definition, progressive government.

In his autobiography, written in 1913, La Follette reviewed the accomplishments of the Railroad Commission's first three years in regulating electric utilities, claiming that it had improved service and had reduced rates by $375,000 per year. At the same time, La Follette asserted, the commission had fostered the growth of utilities, enabling them to increase their investment by 35 per cent. La Follette asked, "How has it been possible that both the people of Wisconsin and the investors in public utilities have been so greatly benefited by this regulation?" He answered the question himself as follows:[72]

> Simply because the regulation is scientific. [*La Follette's emphasis*] *The Railroad Commission has found out through its engineers, accountants, and statisticians what it actually costs to build and operate the road and utilities. On the other hand, since the Commission knows what it costs, it knows eactly the point below which rates cannot be reduced. It even raises rates when they are below cost, including reasonable profit.*
>
> *The people are benefited because they are not now paying profits on inflated capital. The investors are benefited because the commission has all the facts needed to prevent a reduction of rates below a fair profit on their true value.*

Conclusion

Anxious to do something about the public utility problem, encouraged by the examples of New York and Wisconsin, and supported by the favorable opinion of a wide assortment of interests, the legislatures of nearly two thirds of the states passed laws creating comprehensive state commissions to regulate electric power and other utilities in the half-dozen years following 1907. Vermont was first after New York and Wisconsin, and established its commission in 1908; Maryland followed in 1910. The next year nine more states (California, Connecticut, Georgia, Kansas, Nevada, New Hampshire, New Jersey, Oregon, and Washington) enacted laws creating public utility commissions. In 1912 only one state, Arizona, created a state commission, but in the next year new commissions were formed in the following states before the movement was temporarily exhausted: Colorado, Idaho, Illinois, Indiana, Maine, Massachusetts, Missouri, Montana, North Carolina, Ohio, Oklahoma, Pennsylvania, Rhode Island, Virginia, and West Virginia.[73]

The National Civic Federation had advocated regulation by the states because state regulation promised to rectify the unfair treatment given different classes of customers by the utilities and because it would ensure evenhandedness in the setting of rates and the authorizing of stock issues. Best of all, a state commission that was scientific and expert would take public utilities out of politics, thereby removing one of the chief obstacles to good government. Hughes, La Follette, Johnson, Wilson, and other governors sponsored legislation establishing state commissions for similar reasons and also as a means of increasing their power while weakening their opponents, the big-city machine politicians. In the electric power industry, the heads of many—though not all—of the leading utility systems supported the movement for commissions because they believed commissions would protect them from political bosses, from competition, and most of all from municipal ownership.

In short, the concept of state regulation was both compatible with the ideas and political needs of progressives and expedient for safeguarding the material interests of the utilities. From 1907 to 1913, philosophical compatibility and commercial expediency combined to produce a political necessity.[74]

Endnotes

1. William E. Mosher and Finla Crawford, *Public Utility Regulation* (New York: Harper, 1933), p. 551 wrote: "with unvarying consistency and stubbornness, all the political influence at [the utilities'] command has been mustered, in the first instance, to oppose the establishment of regulatory bodies and later the exten-

sion of the powers of such bodies, when once established." *Public Utility Regulation*, p. 551. *See also* Merle Fainsod, Lincoln Gordon, and Joseph C. Palamountain, Jr., *Government and the American Economy*, 3rd ed. (New York: W. W. Norton, 1959), p. 317.

2. Gabriel Kolko argues that railroads sought federal regulation to escape from state regulation. Exactly the opposite was true of electric utilities, which considered state regulation preferable to federal or local control. *See* his *Railroads and Regulations, 1877–1916* (Princeton: Princeton University Press, 1965).

3. McDonald, *Insull* (Chicago: The University of Chicago Press, 1962), p. 55.

4. Ibid., pp. 84–90.

5. Ibid.

6. National Electric Light Association, *Proceedings* (New York: National Electric Light Association, 1898), pp. 14–29.

7. Ibid., pp. 27–28.

8. Ibid., (1905), pp. 6–7.

9. McDonald, *Insull*, p. 117.

10. *Proceedings*, 1 (1907), p. 517.

11. Ibid., Appendix A, p. 10.

12. Ibid., p. 9.

13. Ibid., pp. 20–21.

14. Ibid., p. 12.

15. Ibid., pp. 11–12.

16. Ibid., p. 16.

17. Ibid., p. 16. It is interesting to note that the concept of "automatic regulation" has resurfaced in utility rate cases in the past ten years. Its most widespread use has been in fuel cost adjustment clauses, which have been adopted in nearly every state, but in 1975 the New Mexico Public Service Commission went one step further and adopted a "cost of service index" that allows the Public Service Company of New Mexico to raise its rate automatically to keep up with costs without a hearing before the public utility commission. Such a mechanism is very much like the type of regulatory control proposed by the NELA's Committee on Public Policy in 1907, but today no utility company outside of New Mexico can count on being able to effect regular rate increases without first doing battle with its state regulatory commission. See *Business Week* (September 26, 1977), p. 84.

18. *Proceedings*, 1 (1907), p. 27.

19. Ibid., p. 30.

20. Ibid., p. 4.

21. McDonald, *Let There Be Light* (Madison, Wisconsin: American History Research Center, 1957), p. 118.

22. *Proceedings*, 1 (1908), p. 3.

23. Ibid.

24. Frank Bergen, "Restrictive Legislation Against Public Service Corporations in New Jersey," *Annals of the American Academy of Political and Social Science*, LXXXI (May 1908), pp. 659–670.

25. National Civic Federation, *Municipal and Private Operation of Public Utilities*, vol. 1, pt. 1 (New York: National Civic Federation, 1907), p. 32.

26. Alex Dow, *Some Public Service Papers by Alex Dow, 1892–1927* (Detroit: privately published, 1927), p. 204.
27. Raymond C. Miller, *Kilowatts at Work: A History of the Detroit Edison Company* (Detroit: Wayne State University Press, 1957), pp. 130–131.
28. Ibid.
29. Public Act 1909, Number 106, Sec. 7, cited in Shelby B. Schurtz, "The State Public Service Commission Idea" (Address presented before the convention of The League of Michigan Municipalities, Grand Rapids, Michigan, July 26 and 27, 1917), p. 15, on file at Baker Library, Graduate School of Business Administration, Harvard University. One would be just as much in error to suggest that all civic reformers were in favor of state regulation as to suggest that all heads of public utilities were. In the large cities of Illinois, Ohio, Michigan, and Minnesota, home rule advocates were often "good government" reformers. In 1915 Charles Merriam, professor of political science at the University of Chicago and a one-time candidate for mayor of Chicago, argued:

"It is difficult in any state to make the choice of a utilities commission an effective issue in the selection of a governor, and it is precisely for this reason that public utility interests, as a rule, prefer that type of regulation. Upon this point they are certainly 'wiser in their day and generation than the children of light.' I know that the public service corporations of Chicago will never be as effectively regulated in the public interest by the state, as they will be by the city of Chicago, and I have reason to believe that the same situation is found in many other large cities. In my opinion the cry that 'politics' will interfere with adequate municipal regulation is in itself one of the cleverest pieces of 'politics' in the long history of clever utility corporation tactics."

Charles E. Merriam, "The Case for Home Rule," *Annals of the American Academy of Political and Social Science, LVII* (January 1915), p. 174. The home rule movement was so successful in Minnesota that it was not until 1975 that a state commission gained jurisdiction over electric utility rates. For an early history of the home rule movement in Minnesota, see Stiles P. Jones, "What Certain Cities Have Accomplished Without State Regulation," *Annals of the American Academy of Political and Social Science, LVIII* (January 1915), pp. 72–82.

30. Mansel Griffiths Blackford, "Businessmen and the Regulation of Railroads and Public Utilities in California During the Progressive Era," *Business History Review XLIV* (1970), p. 313.
31. Proceedings of the 1909 annual convention of the Pacific Coast Gas Association, cited in ibid., p. 313.
32. Proceedings of the 1910 annual convention of the Pacific Coast Gas Association, cited in ibid., p. 506.
33. *San Francisco Examiner*, December 8, 1911, cited in ibid., p. 314.
34. *Pacific Gas and Electric Magazine*, March 1912, cited in ibid., p. 315.
35. Samuel Insull, "The Obligations of Monopoly Must Be Accepted," in his *Central Station Electric Service* (Chicago: privately printed, 1915), pp. 119–120.

36. For a history of the NCF, see Gordon Maurice Jensen, "The National Civic Federation: American Business in an Age of Social Change and Social Reform, 1900–1910" (Ph.D. diss., Princeton University, 1956).

37. National Civic Federation, *Municipal and Private Operation of Public Utilities*, vol. 1, part 1 (New York: National Civic Federation, 1907), p. 12.

38. Ibid., p. 23.

39. Ibid.

40. This point was argued by Edgar and Walton Clark, but it found expression elsewhere in the report as well. *See* ibid., pp. 25–26, 40–42, 309.

41. Ibid., pp. 25, 100.

42. Ibid.

43. Ibid.

44. Ibid., p. 40.

45. Ibid., p. 26.

46. Ibid., p. 41.

47. John R. Commons, *Myself* (New York: Macmillan, 1934) pp. 111–120.

48. National Civic Federation, *Draft Bill for the Regulation of Public Utilities* (New York: National Civic Federation, 1914).

49. Schurtz, "State Public Service Commission Idea," p. 30.

50. La Follette was a United States senator in 1907 when the Wisconsin law was passed, but the political contest that cleared the way for its passage had occurred two years earlier while he was governor.

51. *See* Editors' Introduction, in David J. Danelski and Joseph S. Tulchin, eds., *The Autobiographical Notes of Charles Evans Hughes* (Cambridge, Mass.: Harvard University Press, 1973); *see also* Robert F. Wesser, *Charles Evans Hughes: Politics and Reform in New York 1905–1910* (Ithaca, N.Y.: Cornell University Press, 1967).

52. In 1941 John Lord O'Brien, who served as a member of the assembly during Charles Evans Hughes's tenure as governor, wrote:

"It is impossible to reproduce for the present generation the aggressive personality, the seemingly inexhaustible energy, the unrelenting insistence of Governor Hughes. It is equally impossible for those who were not eye witness to appreciate the dramatic qualities which characterize his efforts, or the overpowering quality of his arguments during those stormy years. . . . Without creating a political faction of his own, he brought to New York State an entirely new kind of political leadership, the effects of which still remain potent. His genius for evoking by personal persuasion the support of popular opinion was as unique as it was successful. The numerous statutes enacted on his insistence present, in themselves, a remarkable record of achievement. But over and above these reforms, it was the force of the man's personality, his genius for masterly exposition and the great confidence of the public in his personal integrity which brought to him the support of the electorate and challenged the attention of the Nation at large."

American Bar Association Journal, July 1941, cited in Merlo J. Pusey, *Charles Evans Hughes*, vol. 1 (New York: Macmillan, 1951), p. 209.

53. Mosher and Crawford, *Public Utility Regulation*, p. 23.
54. This account of the gas and electric inquiry is based on Pusey, *Charles Evans Hughes*, p. 23.
55. Danelski and Tulchin, *Autobiographical Notes of Charles Evans Hughes*, p. 120.
56. Ibid.
57. Pusey, *Charles Evans Hughes*, p. 134.
58. *New York Evening Mail*, April 1, 1905, cited in ibid., p. 135.
59. Ibid., p. 231, describing an article that appeared in the *New York World* of May 14, 1908.
60. *New York World*, March 25, 1905, cited in ibid., p. 136.
61. Ibid., pp. 138–139.
62. The report of the Hughes committee, cited in ibid., p. 139.
63. See Henry Bruère, "Public Utility Regulation in New York," and Horatio M. Pollock, "The Public Service Commissions of the State of New York," in *Annals of the American Academy of Political and Social Science*, XXXI (May 1908), p. 542 and p. 653, respectively.
64. Danelski and Tulchin, *Autobiographical Notes of Charles Evans Hughes*, p. 121.
65. Ibid.
66. Ibid., p. 183.
67. Ibid., p. 206.
68. Ibid., p. 143.
69. John R. Commons, "The Wisconsin Public Utilities Law," *Review of Reviews*, XXVI (1907), pp. 221–224.
70. Walter Wellman, "The Rise of La Follette," *Review of Reviews*, XXXI (1905), p. 299.
71. Ibid., p. 300.
72. Robert M. La Follette, *La Follette's Autobiography* (Madison: University of Wisconsin, 1960), p. 153. See, however, Stanley P. Caine's study of Wisconsin railroad regulation, *The Myth of a Progressive Reform* (Madison: The State Historical Society of Wisconsin, 1970).
73. Schurtz, "State Public Service Commission Idea," p. 6.
74. I am indebted on this point to Forrest McDonald, *Let There Be Light*, p. 119.

Chapter 3

REGULATORY TASKS
AND STRUCTURAL CHANGE

In essence, all of the supporters of electric utility regulation shared a common desire—the desire to be protected. Consumers sought protection from high rates; progressives sought protection from political machines and monopoly power; electric utilities sought protection from municipal ownership and from the effects of competition. From its inception, utility regulation has reflected this defensive posture. Its functions have been largely negative—aimed at preventing the worst abuses rather than at promoting the optimal use of economic resources.[1]

This negative orientation has had two important effects. First, it has meant that small consumers of electricity have ignored the regulatory process unless some scandal or political entrepreneur succeeded in making real or imagined abuses highly visible. Second, it has meant that throughout most of the seventy-five year history of electric utility regulation by the states, the utilities themselves have been free to make virtually all of the important decisions regarding the production and marketing of electricity as long as commissions judged their rates to be "fair and reasonable." These commissions, in turn, were for the most part "tightly constrained" by the industry because of the lack of other sources of external support.[2]

In the late 1960's, however, changes in the technology and cost of electric power production altered this state of affairs. New participants in the regulatory process tore away the mantle of obscurity that had cloaked state utility commissions and upset the division of labor between the regulatory commissions and the industry. In the language of Chapter 1, both the stringency of external constraints

61

on regulatory behavior and the nature of regulatory tasks underwent a structural evolution. In this chapter we will examine these structural shifts as a prelude to more detailed case studies of the regulatory response in New York and California. But first, it is important to provide a brief description of traditional regulatory tasks, so that these recent changes may be appreciated.

Regulatory Tasks

State utility commissions and their staffs vary in size, in mode of selection, in jurisdiction, in statutory authority, and in the pay they receive.[3] The complexity of these variations is deceptive. In fact, most state commissions can be classified as either leaders or followers. Almost always, important regulatory innovations are implemented first in the states with the largest staffs and most resources—New York, California, and Wisconsin. Other states adopt a wait-and-see attitude, preferring to observe which innovations are successful in the big states before adopting similar procedures of their own. With time the ratemaking procedures of most states come to resemble closely those of the leaders, despite the wide degree of variation in the institutional details of regulation. One consequence of this phenomenon is that differences among states are most striking during the early stages of regulatory change.

The "mature" regulatory process is comprised of three primary phases: (1) the day-to-day activities of the staff; (2) the formal regulatory process, consisting of a trial-like hearing presided over by the commission or an administrative law judge in which evidence is presented and challenged; and (3) the informal regulatory process, consisting of the interaction of commissioners, their staffs, the regulated utilities, various business and political groups (called intervenors), and the political leadership of the state.[4]

The Day-to-Day Activities of Regulatory Staffs

The day-to-day activities of commission staffs include processing the data on operations and finances submitted at regular intervals by utilities; inspecting meters for accuracy and plant facilities for safety; auditing accounts, either continuously or intermittently; and recording and investigating consumer complaints. For many utility customers the complaint process is the essence of regulation. In-voking the name of the local department of public utilities and threatening to file a complaint is often the only recourse an upset customer has when, for example, the gas company fails to supply service promptly.

Consumer complaints, together with mandatory reports and the results of on-site investigations, serve as the early warning system of the commission. Complaints and spot checks by commission inves-tigators are primarily useful in identifying problems involving service quality or safety, since the rates utilities charge—once ap-proved by the commission—are assumed to be "just and reason-able." In New York and California, utilities' monthly, quarterly, and annual financial reports on operating expenses and revenues are closely examined by commission staffs to track the earnings of the utilities. If a staff thinks certain companies are earning excessive rates of return, it may recommend that the commission institute a formal hearing on its own motion to investigate the level of earnings, or it may recommend that the commission seek to negotiate a reduc-tion in rates informally.

The decision to institute formal rate proceedings is an important one. Commissioners and their staffs like rate reductions. Yet in New York (which is representative) during the ten-year period from 1960 to 1970 when rate reductions were possible, only two formal pro-ceedings aimed at lowering electric rates were instigated. Of the two proceedings, one appears to have been caused more by political pressure from the mayor of Rochester than by the commission's concern over excessive profits.[5]

There are a number of possible reasons for this paucity of commission-instigated proceedings. First, the commission may have felt that only one or two proceedings were necessary to convince other utilities to file voluntary reductions rather than to wait for the commission to order a hearing—a sort of "station the highway cop in full view" strategy. Second, the commission may have felt that a small, quick, and negotiated reduction was preferable to a larger and later one because of the time and resources necessary to pursue a formal proceeding. Because rates, once established, are presumed to be reasonable, the burden of proof is on the party proposing a change in rates. When utilities file for an increase, they have to prepare the affirmative case; if the commission decides to seek a decrease, its staff has to develop the case.

A third—and more likely—reason for the lack of New York commission proceedings is that there was simply no great popular demand for rate reductions during the period from 1960 to 1970. The real price of electricity was falling, so consumers were satisfied. Rather than anger the utilities, the commission decided to avoid controversy and negotiate. It certainly was not overworked in those years. From 1960 to 1967 the New York commission issued only seven opinions in rate cases, compared with twenty-five in the four-year period 1968–1971.[6]

Rate reductions are not common in today's inflationary world; rate increases are. When a utility decides that its rates are too low to allow it to earn a reasonable return on investors' equity, it files a proposed increase in its tariffs, or rates, with the appropriate utility commission. Typically, the new rates are immediately suspended for a period of three to ten months while the commission institutes a formal rate proceeding to determine whether the new rates are "fair."

The Formal Regulatory Process

The elements of the formal regulatory process used to set rates for electric utilities have developed greatly in the past seventy years. The process itself has become known as "cost-of-service ratemaking."[7] There are five key steps in cost-of-service ratemaking: the first four, considered together, determine the total revenue a utility may earn—its "revenue requirement"—and the fifth is devoted to a consideration of how rates will be designed to yield the authorized revenue—"the rate structure." The process of determining the revenue requirement (RR) is easy to describe, as follows:

1. A test year is selected for the utility. Usually this is the most recent twelve-month period for which data exist.
2. The following expense items are calculated for the test year and added.
    (a) Operating Costs (OC)
    (b) Depreciation (D)
    (c) Taxes (T)
3. The utility's total net investment, its rate base (RB), is calculated for the test year.
4. A rate of return (r), which represents what regulators consider to be a reasonable profit for the utility to earn on its

investment, is calculated and the rate base multiplied by it. That figure is added to the amount determined in step two to equal the revenue requirement.

The basic formula for setting rates can thus be expressed:

$$RR = OC + D + T + (RB)r.$$

Once the revenue requirement has been established, the only remaining step is to determine how the rates will be structured—that is, which classes of customers will be charged how much. Until recently this step has been left almost entirely to the discretion of utility management, with the only requirement being that rates not be "unduly discriminatory." Once the rates are approved, it is only the *rates*, not the total revenue or earnings of the utility, that are fixed and may not be changed without commission authorization. As long as a certain tariff schedule is in effect, any excess profits a company earns by charging such rates are not subject to refund.

In part this absence of control over rate structures has been due to a lack of commission expertise, time, and access data, but there are two other, more important, reasons. First, regulators have always conceived their function to be more one of preventing excess profits than of promoting economic efficiency. Consequently whatever resources commissions have had available to them have been devoted to strengthening their ability to challenge the utilities' revenue requirement cases. Second, from the end of World War II until the late 1960's the price of electricity was of no great concern to the average residential consumer. Monthly bills were stable and not large. Indeed, in the two decades from 1951 to 1971, the real price of energy actually dropped 43 per cent.[8] As long as rates were falling, consumers were indifferent about how utilities structured their rates. In the absence of customer complaints, state commissions—even the largest and most skilled ones—had little incentive to exert systematic control over utility rate structures.

From this description of regulatory tasks, it can be seen that regulators were seldom required to engage in much planning or to take much responsibility for the economic effects of utility rate structures. That situation has, however, changed dramatically in the past ten years. State commissions have been challenged as never before to assume responsibility for major decisions regarding both the production and the pricing of electricity. To see these changes in the proper context, it is necessary to understand the economics of electric power production and the general method of setting rates.

Electric Power Production and Pricing

Electric power is subject to a condition that other commodities are not: It cannot be stored. Since electricity must either be used or wasted the moment it is produced, there is no inventory to draw upon when demand is especially heavy. As a consequence, power facilities have to be large enough to meet a system's *peak* demands. Peak periods vary by time of day and by season. From an efficiency standpoint, customers who use power "on-peak" should pay the added or marginal cost of providing the extra capacity needed to serve them. "Off-peak" users—customers who use power when much of the system's capacity is idle—should not be charged at all for power plant capacity but only for the costs of the fuel, labor, and materials required for the satisfaction of their demands. Since the investment in electric plants is large and plants' overhead costs are high, the difference between what is charged for electricity taken on-peak and that which is taken off-peak can be substantial.

Contrary to what is indicated by economic theory, the rates charged by most electric utilities take into account the volume, *not* the time of use. The pattern of these rates is often referred to as a "declining block tariff" or a "promotional rate design" because the cost per kilowatt-hour of electricity declines with increased usage. Large users may pay only half as much per kilowatt-hour as small users.[9] Promotional rates were instituted in this country largely through the efforts of Samuel Insull, who first learned of their use in England in 1894.[10] During the development of the industry, when economies of scale could be realized by the construction of larger and larger plants, promotional rates made a good deal of sense. By promoting use, larger plants were built and the price per kilowatt-hour was lowered. With the maturing of the industry, such opportunities were depleted and promotional rates became more difficult to justify. Utilities which still cling to a system that gives lower rates to industrial and large-volume users do so in large part because they are discriminating in price in accordance with elasticity of demand, not because they are fostering efficiency.

It is generally conceded that industrial users have a more elastic demand for electricity generated by utilities than do residential customers because they have the option of generating their own power. Residential customers, meanwhile, have a greater elasticity of demand for electric space heating than for lighting. Instead of electricity, householders can use oil or gas to heat their homes; few, however, would use candles or gas lamps in place of electric lighting. By providing a quantity discount, utilities can give a lower price to customers that are more likely to abandon electricity or generate their own electricity when its price rises.

In setting their rates, utilities divide their costs into three categories: output costs, customer costs, and demand (or capacity) costs.[11] Output costs include the costs of fuel, labor, and materials—the short-run variable costs of production. Customer costs are incurred in such activities as reading meters and billing accounts. Demand charges refer to the fixed costs of plant operation and management necessary to provide enough power to meet a system's peak demand.

The proper method for allocating demand charges is a topic on which economists and utility accountants disagree strongly. The utilities consider demand costs to be "readiness to serve" charges, measurable by a customer's maximum consumption during a given past period. Economists maintain that from the point of view of efficiency, no demand costs should be assessed against any customer for power taken during the off-peak period, because a kilowatt-hour of usage off-peak does not require any additional plant capacity. They argue that costs can be lowered not by promoting use at any time but by increasing a system's "load factor." A system's load factor is the ratio of its average load to its peak load; the higher the load factor, the less idle capacity in the system. According to economists, load factors can be improved by encouraging customers to increase their consumption of power off-peak. Declining block rates fail to provide this incentive because they do not discriminate by time of use. In response, utilities argue that promotional rates are justified because they recover demand and customer charges in the first blocks—low-volume increments of usage—and charge only for the lower operational costs in the tail blocks—high-volume increments. However, high-volume use, if taken on-peak, adds not only to operational costs but to capacity costs because a larger plant is required to meet peak demands.[12]

Although economists have long talked about and written about the importance of restructuring electric utility rates so that they are in

greater conformity with marginal costs, few of them had much hope
that their advice would be acted on. In 1966, for example, William
G. Shepherd, a distinguished student of utility regulation, concluded:[13]

> *Although rate-structure problems will probably get more regulatory
> attention than before, it would be rash to expect any important im-
> provements in those existing rate patterns that obviously violate allo-
> cational criteria. For example,* no change from declining-block elec-
> tricity pricing—which gives almost exactly the wrong relative pricing
> for peak-demand use—is in sight. [*emphasis added*] *Indeed, it is
> difficult to imagine anything less likely than that economic analysis of
> allocation would be incorporated into regulation at the state-
> commission level. It is not in our time that the National Association of
> Railroad and Utility Commissioners* [National Association of Regula-
> tory Utility Commissioners] *will be heard to stress, for example, the
> marginal conditions of efficient investment allocation in the utility
> sector, as part of the national economy.*

Yet precisely those things Shepherd and others said would not
happen are happening in the leading states, and they will likely
happen elsewhere as well. The reason for this new interest in sub-
jecting rate structures to closer regulatory supervision is, as Paul
Joskow says, not that utility commissions have suddenly "seen the
light." Instead, their changed behavior is a response to alterations in
the economic and political environment in which they operate.[14]

Structural Change in the Political Economy
of Regulation

From the end of World War II until the late 1960's, promotional
campaigns urged people to "live better electrically" by purchasing
electric appliances and building "all electric homes." The declining
block rate structure that utilities used to price electricity stimulated
growth in demand. The more consumers used, the less per unit
they paid. Everyone—utility executives, investors, state regulators,
and consumers—seemed satisfied with promotional pricing because
the growth in demand meant that larger generating plants could be
constructed and greater economies of scale realized. Investors
liked the larger plants because they promised increased profits.
Utility managers were attracted to the opportunities for growth in
sales. Regulators had the happy task of watching the industry become more efficient and, on occasion, of negotiating rate reductions.

In the absence of conflict, newspapers ignored the utilities except for an occasional story on the financial page noting the issuance of bonds to finance new construction. Planning and pricing decisions—how many new plants should be built, what type of plants should be built, where and when they should be built, and who should be charged for them—were largely made by the industry itself with little supervision by government regulators. The chairman of one of New England's commissions put the point even more emphatically: "Regulation during the 1960's," he said, "was non-existent."

Economic Change: Inflation and the Increase in Formal Regulation

As long as rates were stable or declining and service was reliable, about the only people who paid any attention to state utility commissions were utility executives. In most instances, this meant that commissions were in the backwater of state politics. Appointments were often political payoffs which largely went unnoticed. It was not uncommon for commissioners to treat their office as "no-show" jobs. Many were poorly trained and were not at all interested in the technical affairs of utility regulation.

As one long-time observer of state commissions remarked: "You can put your brother-in-law who needs a job in as earthquake commissioner . . . and he'll do fine—as long as there is no earthquake." For many state commissions, the earthquake hit in October 1973 in the form of the Arab oil embargo. As a direct result of the embargo, fuel costs skyrocketed to nearly four times their 1972 levels. These costs were soon reflected in increased electricity rates. Nationwide, rates rose 90 per cent in the five years after 1970. In New York the increase was even sharper: some customers were paying nearly two times as much per kilowatt-hour in 1974 as they had in 1972.[15] People who had been induced to buy "all electric" homes were especially hard hit. Consolidated Edison, New York's giant utility, reported that the typical monthly bill for residential electric heat in Westchester County was about $250 in March, 1974, which was up from around $130 one year earlier.[16] Some Georgia Power customers even went so far as to sue the electric utility, claiming that the company's promotional advertisements that totally electric homes were efficient were known by it to be false.[17]

But rising fuel costs were only one source of the rise in electricity rates. By 1967 the economies of scale and other technological advances which had caused electricity prices to fall from World War II levels had largely been achieved.[18] The effect of inflation in the cost of capital and in construction wages soon swamped the savings associated with larger size. From 1972 to 1975 the cost per kilowatt of new nuclear capacity rose 80 per cent while the cost of new coal-fired power plants doubled.

These dramatic increases seriously threatened the financial position of the utilities, which deluged regulatory authorities with requests for permission to raise rates. The increase in regulatory activity is reflected in the number of general rate reviews conducted by state utility commissions. In 1963 only 3 cases were being reviewed nationwide. By 1969 the number had increased to 19, and by 1975 it had shot up to 114 (see Table 3.1). Utility commissions had neither the staff nor the administrative capability to cope with this huge new demand on the formal regulatory process.

From the point of view of the utilities, the problem with general rate cases is that they take so long to be decided. In some of the larger cases in the mid-1970's, it was not unusual for the process to take eighteen months or even longer.[19] To counteract the effect of

Table 3.1   Formal Rate of Return Reviews Processed by State Regulatory Commissions* 1963–1975

Year	General Rate Reviews	Average Cost cents/kwh	Average Revenue cents/kwh
1963	3	1.33	1.77
1964	4	1.30	1.73
1965	2	1.27	1.70
1966	5	1.24	1.67
1967	3	1.24	1.66
1968	8	1.24	1.64
1969	19	1.22	1.63
1970	45	1.25	1.68
1971	51	1.34	1.78
1972	94	1.40	1.86
1973	64	1.49	1.97
1974	78	1.96	2.50
1975	114	2.30	2.94

Source: Paul L. Joskow, "Electric Utility Rate Structures in the United States" (Paper presented at the Seventh Michigan Conference on Public Utility Economics.) Reprinted by permission of Westview Press from *Public Utility Rate Making in an Energy-Conscious Environment*, edited by Werner Sichel. Copyright © 1979 by Westview Press, Boulder, Colorado.
* Average cost and revenue data in current dollars.

"regulatory lag," it became common for utilities to file a new request for permission to raise rates immediately upon completion of a previous rate case and to urge regulators to adopt or to extend the use of "fuel cost adjustment clauses" which allowed them to increase rates to cover escalating fuel costs without first going through a lengthy hearing process.

Helpful as the fuel adjustment clause was to the utilities, it did not solve the companies' financing problems altogether. A staff member of one Northeastern state utility commission recalled: "When the embargo hit, it had a tremendous impact on the reliability of service and on electric rates. The crisis caused economic chaos. The companies couldn't increase their working capital fast enough to meet the huge increase in fuel costs. Even with the fuel cost adjustment clause there is still a lag, and they couldn't get the revenues they needed. The companies had to pay out to buy their fuel before they could get their revenues back."

In one case in particular, that of Consolidated Edison in New York, the 1973 oil embargo hit with such fury that even the fuel clause was unable to save it from near financial disaster. Under strict state and local anti-pollution laws, Con Ed had converted many of its coal-fired plants to oil. Its nuclear plant construction project had been effectively stymied by environmental groups. Consequently in 1974 fully 75 per cent of its electricity was produced by burning foreign oil.[20] It was the wrong time to be burning imported oil. An engineer at the New York Public Service Commission recalled the company's plight:

> Con Ed was immediately affected by the embargo . . . A black market for oil developed with Con Ed buying oil on the high seas at $23 to $24 per barrel—up from $4 to $5 per barrel earlier. . . . Even though the New York State Public Service Commission authorized a temporary increase of something like $75 million in February 1974, in April the company was on the brink of bankruptcy because it couldn't raise working capital. . . . Working capital is maintained on hand to tide them over between outflow for fuel costs and operating expenses and wages, and inflow of revenue. . . . Banks were refusing to loan and people weren't buying its bonds. When the commission refused to allow the company to use working capital to pay [common stock] dividends, the company had to pass on its dividends in April. The stock stopped trading for awhile.

It was the first time in eighty-nine years that the giant utility or its predecessors had failed to pay quarterly dividends on its common

Table 3.2   **Average Yields of Electric Utility Debt and Earned Returns on Common Equity, 1963–1975**

Year	Yield on Electric Utility Debt	Rate of Return on Common Equity
1963	4.40	11.4
1964	4.55	11.8
1965	4.61	12.2
1966	5.53	12.4
1967	6.07	12.4
1968	6.80	11.9
1969	7.98	11.8
1970	8.79	11.2
1971	7.72	11.0
1972	7.50	11.1
1973	7.91	10.8
1974	9.59	10.2
1975	9.97	10.6

Source: Paul L. Joskow, "Electric Utility Rate Structures in the United States" (Paper presented at the Seventh Michigan Conference on Public Utility Economics.) Reprinted by permission of Westview Press from *Public Utility Rate Making in an Energy-Conscious Environment*, edited by Werner Sichel. Copyright © 1979 by Westview Press, Boulder, Colorado.

stock. In a desperate attempt to raise cash, the utility's chairman, Charles F. Luce, proposed that the Power Authority of the State of New York (PASNY) buy two of Con Ed's unfinished generating plants for $450 million.

The shock of Con Ed's actions reverberated throughout the industry. The common stock of Boston Edison fell from over $25 on April 22 to about $15 on May 14. Duke Power's stock dropped over 12 per cent. Financial analysts worried about the industry's ability to raise capital. Said one: "Con Ed lost the institutions long ago, but its dividend allowed it to count on the little old lady in tennis shoes. Now it has lost her, too. And lost her for a lot of other utilities." Said another: "Utilities . . . have long been considered the best stocks for widows and orphans, and now we have the largest of them in trouble and passing its dividend. This throws the viability of the entire industry in doubt. It makes it more difficult for us to sell securities at a time when we need to sell them more than ever." Some executives began reexamining ambitious construction plans in light of the new fuel and capital costs. Toledo Edison cut a stock offering in half, while Detroit Edison hesitated before offering an issue of $150 million and then slashed it by a third to get it sold. The American Electric Power Company (AEP) reported its intention to make drastic cutbacks in its construction program—at least $250 million in two years.[21] As profits lagged, financial analysts downgraded the bonds of many electric utilities, which meant that the utilities and ultimately the consumer would have to pay more to borrow money for new construction (see Table 3.2). From 1974 through June 1977 there were 184 changes in the ratings of electric utility debt by Moody's and Standard and Poor's, the major rating services; 35 issues were upgraded and 150 downgraded.[22] It soon became common for utility executives like Lelan Sillin Jr., the chairman and president of Northeast Utilities, the largest power supplier in Connecticut, to contend that the regulation of electric utilities had "broken down."[23]

Political Change: Material and Ideological Confrontations

Utility executives were not the only ones to question the efficacy of state regulation. For different reasons, three other groups—consumers, environmentalists, and experts from the Federal Energy Administration (FEA)—began appearing before state regulatory commissions in utility rate proceedings, and contributed importantly to the changing regulatory environment.

Consumer Response. The enormous rate hikes and fuel cost increases that were authorized to preserve the utilities' financial standing soon shocked consumers out of their state of indifference. Organizations mushroomed across the country as irate consumers crowded into the once-sleepy hearing rooms of state public utility commissions to voice their opposition to further rate increases. In close pursuit of the consumers came television and newspaper reporters, ever eager to broadcast a confrontation. Regulators, who a few years before had spent their days in relative obscurity debating such arcane matters as the proper valuation of a utility's rate base and the correct treatment of depreciation, now saw the debates recast in emotional terms before a wide audience. Instead of simply addressing the question of whether certain tax credits should be "normalized" or "flowed through" in the usual course of receiving expert testimony and cross-examining witnesses at a commission hearing, regulators became used to reading in the newspapers the charges of some consumer group that the commission had allowed utilities to collect "phantom taxes" and being asked to respond to the charges by television interviewers.[24]

Commissioners were caught in a double bind: If they failed to grant rate requests, the utilities would threaten that they were faced with insolvency, at worst, or much higher prices for new capital, at best. On the other hand, approval of rate increases made the commission appear to be rubberstamping industry requests, and that resulted in more and more consumer representatives appearing at rate hearings, which in turn attracted even more attention from the press.[25]

Environmental Representation.   Coinciding with the awakened interest of consumers, a new and more ideologically motivated group of intervenors began showing up at commission hearings in the early seventies—the environmentalists. Groups such as the Sierra Club and the Environmental Defense Fund (EDF) identified the electric power industry as a major source of environmental degradation, its pro-growth ethos and pricing policies being especially suspect. Where the industry's engineers saw modern, technically efficient generating units, environmentalists saw foul air, scarred landscape, and polluted streams. They argued that utilities failed to internalize the social costs of their actions into their pricing structure and so encouraged consumers to use more electricity than was optimal. Moreover, because the price of electricity did not vary by time of use, large customers were forcing utilities to build more power plants than were needed to meet their peak demands. Starting in 1973 the EDF and other groups began making appearances in rate cases in Wisconsin and then in Michigan, New York, and California to urge regulators to pay more attention to the environmental effects of the industry's building and pricing plans.

As a result of the entry of these two new participants, the scope of conflict over the regulation of electric utilities widened dramatically. Decision-making no longer was the sole province of the industry with the mild supervision of the regulators. A host of new regulations governing the construction and siting of power plants was promulgated by state legislatures and utility commissions. Politicians, the press, and the environmental and consumer groups all clamored for a role.

The interests of the new participants were by no means identical. Environmentalists wanted to stop pollution; consumers wanted to stop rate increases. At about the same time environmentalists discovered they could not stop all new power plant construction, consumers found themselves unable to prevent all rate increases. The

formal hearing process strongly favors those who can present a positive alternative to a proposal or who have the resources to cast
serious doubt on a proposal. In 1972 environmentalists and consumers could do neither. They objected to new construction or new rate
hikes but had no effective means to counter the utilities' presen-
tations. Each group searched for an opening that would give it access
to industry decision-making before plans were submitted to regulatory authorities. The groups' search soon focused on the neglected
rate structures of the regulated firms.

For both environmentalists and consumers the reform of rate
structures was a way to achieve indirectly what they might fail to
achieve directly. Both saw the key element of utility rate structures,
the declining block method of pricing, as inimical to their interests.
Environmentalists were convinced that under another pricing
scheme—one that did not promote high-volume usage—there
would be less need to build new generating facilities. Consumers
opposed the declining block structure because it appeared to favor
the large user, especially industrial and commercial users, at the
expense of the smaller residential consumer. Together the groups
urged utility regulators to break with tradition and hold "generic"
hearings on rate structure reform and to force utilities to abandon
the declining block method of pricing.[26]

The Energy Crisis. Environmental advocates and consumer
representatives were the first but not the only new participants in
the regulatory process to seek the reform of utility rate structures. In
the executive branch and in Congress, various bureaucratic and
political elements targeted the pricing practices of electric utilities as
a cause of and therefore at least a partial cure for the "energy crisis"
and the deteriorating trade balance that followed OPEC's successful
embargo. These elements' interest in electric utilities derived from
the fact that in 1975 energy consumed in generating electric power
accounted for 26 per cent of the nation's total energy consumption;
from the fact that energy consumption in the electric industry was
growing at a rate twice that of energy consumption in general; and
from their projections of the huge amounts of capital the industry
would require to meet the electricity demands of the next ten
years.[27]

The Federal Energy Administration (FEA) became the first of the
federal elements to urge rate reform as a means of improving utility
load factors, conserving energy, and reducing capital requirements.

In 1974 FEA experts began joining consumer representatives and environmentalists in attending state generic hearings and rate cases in which questions of utility rate structure were being considered. To support its claims, the FEA funded an electric utility rate demonstration program to demonstrate to utilities and regulators "the viability and customer acceptance of innovative electric rates" and to gather and disseminate empirical data on the effects of such rates on electricity consumption patterns.[28] Projects were undertaken in 14 states and Puerto Rico. Based on the projects' preliminary results, the Department of Energy estimated that an improvement of 5 per cent in utility load factors effected through rate reform could, by 1985, save: "250,000 barrels of oil and gas per day; 50,000 megawatts of generating capacity; at least $13 billion net in utility costs."[29]

The FEA's action was followed up in Congress by Representative John D. Dingell (D-Michigan), who in March 1976 introduced a broad-gauged bill to restructure electric utility rates "in order to conserve energy, minimize the need for new generating capacity, and assure reasonable rates to consumers."[30] By the time Dingell introduced his measure, Wisconsin, California, and New York had already acted on rate structure reform, so his bill was aimed primarily at putting pressure on the rest of the states to follow the leaders.[31]

Proposals for Rate Reform

Of the many proposals advanced to reform rate structures, the two that have received the most attention are those for "peak load" pricing and "lifeline" rates. Environmentalists prefer peak load pricing based on marginal costs, which differentiates rates according to time of use. Residential consumer advocates, on the other hand, favor lifeline rates, which price the first several hundred kilowatt-hours of electricity below cost. Logically, the two systems differ fundamentally, although in practice it is possible to combine elements of both.

In addition to environmentalists, peak-load pricing advocates include Federal Energy Administration experts and a number of leading academic and consulting economists who have appeared before state commissions on behalf of the concept. They base their arguments on economic theory: that society's resources will be optimally allocated only if rates reflect the true marginal costs of production. Their appeal, therefore, is for a recognition of the value of efficiency, although the environmentalists promote efficiency not for its own sake but rather because they feel it will raise prices, discourage demand, and prevent the growth of the industry.

Residential consumer groups are willing to allow prices to rise only if the higher prices are to be paid by someone else, such as the utilities' industrial and commercial users. To this end they advocate inverting the rate structure, causing the cost per kilowatt-hour to rise rather than fall with increased volume. Usually coupled with this proposal is another proposal that a "subsistence" quantity of electricity (usually 250 or 300 kilowatt-hours per month) be provided at a subsidized rate—a "lifeline" guarantee. Unlike peak load pricing, there is no consistent theory to justify lifeline rates, although there are, of course, many claims put forward on behalf of the proposal which will be discussed and evaluated in a later chapter. In essence, lifeline rates seek to redistribute the burdens of rising electricity rates from one class to another; consequently they are most appealing to those residential consumers who feel that they have historically paid an unfair share of utilities' costs and to welfare advocates who support lifeline for the poor as a matter of equity.

Conclusion: Changing Environment, Changing Tasks

Both of these proposals emanated from outside of state regulatory commissions. Where they have received serious consideration, however, they have succeeded primarily because regulators saw in them the means for coping with the competing demands that they "do something" about electric rates, the environment, and the energy crisis.

These developments can be summarized in terms of our model of the political economy of regulation developed in the first chapter. Inflation in the cost of producing electricity and increased concern about the effects of electric power production on the physical environment changed the regulatory environment of state utility commissions in the following way. First, as a direct consequence of inflation there was a rapid increase in the number of rate cases that needed to be disposed of through the formal regulatory process. This augmented the obligations of regulatory commissions but did

little to expand their stock of internal resources. The commissions'
need to simplify tasks and to "buy" external support from utilities led
them to adopt fuel cost adjustment clauses. Second, these devices,
in turn, allowed the utilities to circumvent the lengthy hearing process and to pass along cost increases immediately to consumers—the
result of which was increasing rates of unprecedented magnitude.
Third, this enormous surge in electricity prices generated widespread interest in the industry and in the regulatory commissions,
which made regulatory instruments like the fuel cost adjustment
clauses highly visible and therefore less acceptable. These three
developments all affected the demand side of our model. Fourth,
the appearance in agency proceedings of consumer representatives,
environmental groups, and Federal Energy Administration experts
that resulted from these changes provided the beleaguered commissions with new informational, political, and polemical resources at
the same time it placed new demands on them. This affected the
supply side of our model.

These shifts in demand and supply conditions altered agency autonomy zones and the power of utilities to affect agency decision-
making. In one sense state commissions became less dependent
upon the utilities they regulated because they now had other client
groups to whom they could turn for the resources that they needed
to maintain themselves. But in another sense, the new regulatory
environment enhanced the position of the utilities. To a large extent,
as we have seen, the workload of a state utility commission is determined by industry-initiated requests for rate hearings. By controlling the demands on the formal regulatory process, utilities could,
conceivably, affect the degree to which the commissions would need
their cooperation. This, in turn, would put the utilities in a better
position to require concessions in regulatory policy. Whether or not
this actually occurs and the zone of autonomy within which an
agency develops is constricted or widened will depend on whether
the supply effects of increased client variety outweigh the demand
effects of increased dependence on utility cooperation.

These environmental shifts in the process of regulation were accompanied by changes in the tasks of regulation. The two reform
proposals that emerged out of the debate over the proper structure
of electric rates, lifeline rates and peak load pricing, confronted the
state commissions with fundamentally different and new tasks.
Lifeline rates are explicitly redistributive. They have the effect of

shifting the burden of meeting a utility's revenue requirement from one customer class to another. Alternatively, peak load pricing represents an effort to persuade utility commissions to undertake responsibility for promoting the optimal allocation of society's resources.

It can be argued that regulation has always had a distributive, or redistributive, effect and therefore that lifeline proposals are nothing new. But one can accept the notion of the distributive effects of electric utility regulation and still recognize lifeline rates as a fundamental departure from the standard operating procedures of state commissions. In the period after World War II until the early 1970's, as we observed, regulatory commissions did not engage in efforts to structure utility rates. That task was left to the utilities themselves. That the commissions implicitly accepted the distributional effects of these rate structures is true, but there was no attempt to make the design of rate structures a routine task of agency staffs. The same point applies to the effort to use rate structures to promote economic efficiency. Although some early advocates of state regulation did attempt to justify regulation as a "substitute for competition," we have seen that regulation has in practice served a negative, preventative function, rather than a positive, allocative function aimed at promoting the best use of economic resources.

What is the nature of these new tasks and how are they likely to affect the regulatory behavior of state utility commissions? Lifeline rates present commissions with the task of making what we called in Chapter 1 a "point decision." Other than in its initial design, a proposed redistributional rate like lifeline requires nothing more of regulating authorities than a yes/no vote. The proposals to adopt peak load pricing, however, are likely to present commissions with a more complex set of tasks. Since their function is to promote economic efficiency, the rates will require continuous monitoring, evaluation, appraisal, and adjustment as the technical conditions of electric power production shift. This effort will require a good deal of coordination between agency staffs and the regulated utilities, and this means that commissions which accept the logic of peak load pricing are undertaking what we called "planning tasks."

These changes in the regulatory environment and in the nature of regulatory tasks have important implications for the behavior of state utility commissions. If the supply-side effects that were identified predominate, then we would expect the external constraint exerted by electric utilities to loosen and the commissions to be freer to adopt policies independent of industry influence. With lifeline confronting commissions with point decisions and peak load pricing confronting them with planning tasks, this would mean that an investigation of rate structure reform would be delving into Type III and Type IV regulatory situations, respectively, to use the terminology of Chapter 1. If this were the case we would expect that a commissioner's vote in the Type III lifeline case would reflect largely his desire to please constituents in either the "industrial" or the "political" market. If a commissioner chooses to appeal to the political market, we may well see the issue used to generate controversy and visibility in order to enhance his reputation as a "consumer" advocate.

In the Type IV case of developing and implementing a rate structure based on peak load pricing, however, we would expect that a commissioner's desire to maintain his organization—to induce the cooperation and loyalty of his staff—and to secure the cooperation of utilities upon whom he may depend for critical technical information would have a different effect on regulatory behavior. In this case it is likely that a commissioner would appeal to the professional and purposive goals of the agency's staff and would utilize his network of informal contacts to persuade the staff to go along with him. Should there be any disagreement between the commission staff and the utilities, regulatory executives would likely try to minimize conflict and the organizational "strain" that accompanies it. Unlike the Type III lifeline situation where the possibility of conflict escalation exists to serve the political purposes of regulators, we would expect regulators engaged in designing and implementing an economically efficient rate structure to avoid confrontations in order to sustain cooperative efforts within the agency and between the agency staff and the utilities.

The preceding two paragraphs have been written, not unintentionally, in the subjunctive voice in order to indicate conditionality. That is because the changes in the environment and in the tasks of regulation that have been described in this chapter refer to global, or general, changes throughout the country. Specific changes both in demand and supply conditions will vary from state to state. Careful case study is needed to sort out fully all of these influences on the actual behavior of regulators. In some states supply-side effects have predominated; in other states demand conditions have more heavily influenced commission behavior. For example, some utilities, like those in the Northeast, were especially hard-hit by the OPEC oil embargo and subsequent price increase because they generated much of their electricity by burning imported oil. Other utility systems that depended less on oil and more on hydro-electric power or coal to produce electricity were not as dramatically affected. This is a partial explanation for why environmental groups were not as active or as effective in some states as in others.

Because the external conditions affecting agency decisions vary from state to state, it should not be too surprising that state commissions have responded in different ways to the new demands for redistribution and efficiency in rate structures (Table 3.3). Roughly half of the state commissions that adopted lifeline rates by 1977 also adopted time-of-day rates. But over 80 per cent of those commissions that adopted time-of-day rates have rejected lifeline. Many states have shown a strong desire to avoid acting on the proposals for as long as possible, preferring to wait and see what happens in those leading states where the new environmental and consumer interests are most active. As recently as October 1977, only 5,000 customers were subject to time-of-day rates nationwide; only seven states had lifeline rates in effect.[32]

The two largest states, California and New York, have also been the states (in addition to Wisconsin) in which reform proposals have made the most headway. The two states are alike in that their respective public service commissions have large and well-trained staffs that regulate some of the largest investor-owned utilities in the nation. Yet their utility regulators have taken different approaches in dealing with the new demands for public accountability and responsiveness in the pricing of electricity. California has adopted both lifeline and time-of-day rates; New York has eschewed lifeline in favor of peak load pricing.[33]

In the two case studies that follow we shall utilize the conceptual framework developed in Chapter 1 to evaluate the determinants of regulatory behavior in New York and in California; to explain why one proposal was adopted in one state, and not in another; and to describe how programs were developed and put into effect once a particular policy was chosen. It will not be giving away too much of the story of these two case studies to say that each of the commissions encountered a different political environment (in part inherited and in part the result of its own making); that, in consequence, the external constraints on commission behavior differed; and that, to the extent that commissioners were able to exercise discretion

Table 3.3   Status of Electric Rate Structure Reform, 1977

State	Generic Rate Hearings	FEA-Funded Experiments	Lifeline or Inverted Rates	Time-of-Day Rates
Alabama				
Alaska				
Arizona	*	*	*	
Arkansas		*		*
California	*	*	*	*
Colorado	*			
Connecticut	*	*		*
Delaware				
District of Columbia	*		*	
Florida	*		*	*
Georgia				*
Hawaii	*			
Idaho			*	
Illinois	*			*
Indiana				*
Iowa				
Kansas				
Kentucky				
Louisiana				
Maine				
Maryland	*			
Massachusetts	*			*
Michigan		*	*	*
Minnesota				
Mississippi				*
Missouri				

State	Generic Rate Hearings	FEA-Funded Experiments	Lifeline or Inverted Rates	Time-of-Day Rates
Montana				*
Nebraska				
Nevada				*
New Hampshire	*			*
New Jersey		*		*
New Mexico				*
New York	*	*		*
North Carolina	*	*		*
North Dakota				*
Ohio		*		*
Oklahoma		*		
Oregon				
Pennsylvania	*		*	*
Rhode Island		*		*
South Carolina				
South Dakota				*
Tennessee				
Texas				
Utah				
Vermont		*		*
Virginia	*			*
Washington		*		
West Virginia				*
Wisconsin	*	*		*
Wyoming				

Sources: National Economic Research Associates; and Federal Energy Administration, *Electric Utility Rate Design Proposals* (Washington, D.C.: U.S. Government Printing Office, February 1977); and Table 61b, *1977 Annual Report on Utility and Carrier Regulation*, National Association of Regulatory Utility Commissioners (Washington, D.C., 1978), cited in Martin Taschjian and James Hewitt, "State Regulation of Electric and Gas Utilities," *Energy Policy Study*, vol. 4 (Washington, D.C.: U.S. Department of Energy, January 1980), p. 52.

within an expanded zone of autonomy, the fundamental task they set for themselves determined the degree to which bureaucratic factors shaped the content of public policy.

Endnotes

1. Despite the negative orientation of electric utility regulation, its supporters have been fond of arguing that it serves to correct the failings of competition. For an early conceptualization of regulation as a substitute for competition see Alex Dow, *Some Public Service Papers, 1892–1927* (Detroit, privately published in 1927), p. 204.

2. An exception to this generalization is the period from 1925 to 1940, one of the most interesting periods in the history of public utility regulation. Electric utilities fell from favor in the late twenties largely because of a damaging investigation into the financing of the 1926 Senate campaigns in Illinois and Pennsylvania. The investigation was conducted by Senator James Reed (D-Missouri), who saw in the investigation an opportunity to advance his presidential ambitions. In Illinois, Samuel Insull admitted giving $150,000 of his companies' money to Frank L. Smith, head of the Illinois Commerce Commission, which regulated utilities in the state. Smith defeated Senator William B. McKinley, Insull's business enemy, in the Republican primary and later won the general election but was not seated by the Senate. In Pennsylvania, Governor Gifford Pinchot charged that "millions" had been spent by the utilities to defeat him in his bid for a Senate seat.

   The investigation of these campaign financing charges, which were in part documented, was conducted at the time when Congress was considering the disposal of the government's nitrate and power plant at Muscle Shoals, Alabama, on the Tennessee River. Senator George Norris of Nebraska and other progressive midwestern senators were in favor of continued and expanded government ownership of power facilities on the Tennessee River but faced stiff opposition from the utilities. Senator Thomas J. Walsh, the inquisitor of Teapot Dome fame, seized the occasion of the Reed report to move for a comprehensive senatorial investigation into the "power trust." *See* Forrest McDonald, *Insull* (Chicago: University of Chicago Press, 1962), pp. 262–263.

   An exhaustive investigation of the utilities was conducted by the Federal Trade Commission, not the Senate. It resulted in a series of revelations damaging to the industry regarding its use of advertising, newspapers, civic groups, and educational institutions to disseminate propaganda favorable to its point of view and its use of the holding company (a sort of financial pyramid) to avoid regulatory control. *See* Ernest Gruening, *The Public Pays* (New York: The Vanguard Press, 1931) and Carl D. Thompson, *Confessions of the Power Trust* (New York: E. P. Dutton, 1932).

   State after state adopted laws to restrict the utilities' power to influence public opinion, with the result that today advertising expenses and charitable contributions, if they are allowed at all, are among the most carefully scrutinized of utility outlays. In Congress, the Public Utility Holding Company

Act, which some authorities consider to be the most stringent corrective legislation ever applied to American business, was passed and signed into law to prevent some of the more serious financial abuses utilities had committed. *See* Charles F. Phillips, Jr., *The Economics of Regulation*, rev. ed. (Homewood, Ill.: Richard D. Irwin, 1969), p. 564; and Clair Wilcox and William G. Shepherd, *Public Policies Toward Business*, 5th ed. (Homewood, Ill.: Richard D. Irwin, 1975), p. 401.

Earlier, in New York, Governor Franklin D. Roosevelt had entered into a debate with Martin J. Insull, Samuel Insull's younger brother, over the proper role of government regulation of business which was carried in the pages of *The Forum* and the *Proceedings* of the Academy of Political Science. *See* Franklin D. Roosevelt, "The Real Meaning of the Power Problem," *The Forum, LXXXII* (December 1929), pp. 327–332; and "The Revision of the Public Service Commissions Law," *Academy of Political Science Proceedings, XIV* (May 1930), pp. 81–89. Roosevelt argued that utility commissions had strayed from their original mission, which he said was to be the people's advocate: "The Public Service Commission was created in the days of Governor Hughes not to act as a court between the public on one side and the utility commission on the other but to act definitely and directly for the public, as the representative of the public and the Legislature, their sole function being to supervise the utilities themselves under definite rules." Roosevelt, "Real Meaning of the Power Problem," p. 202.

Roosevelt's solution as governor and later as president was to strengthen the supervisory powers of regulatory commissions but not to rely on them as planning agencies. With those ideas in mind he created the Power Authority of the State of New York (PASNY) and later the Tennessee Valley Authority (TVA) to serve as a "yardstick" for private utility companies. Rational planning in the industry would be prompted by the TVA's competitive example. *See* Richard Hellman, *Government Competition in the Electric Utility Industry* (New York: Praeger, 1972), pp. 18–38; and Thomas K. McCraw, *TVA and the Power Fight, 1933–1939* (Philadelphia, Pa.: J. B. Lippincott, 1971).

Roosevelt and the electric power industry were constantly at odds throughout his presidency. In his 1932 campaign he inveighed against holding companies and "the Ishmaels and the Insulls" (McDonald, *Insull*, p. 310). The industry tried to jab back in 1940 when Wendell Wilkie, the president of a utility that had had to sell out to the TVA, led the forces of protest against Roosevelt and the New Deal. *See* V. O. Key, Jr., *The Responsible Electorate: Rationality in Presidential Voting, 1936–1960* (Cambridge, Mass.: Harvard University Press, 1966), pp. 29–62. Forrest McDonald, *Insull*, pp. 272–273, wrote:

"The electric utilities viewed Roosevelt's public power projects as 'creeping socialism' that would ultimately destroy them. To explain Roosevelt's political behavior some industry leaders resorted to *ad hominem* attacks. The most widespread explanation was that early in 1928, before he had decided to run for governor, Roosevelt had approached an old friend, Howard Hopson, head of the multibillion-dollar Associated Gas and Electric Company system, to ask for an executive job. The story went that Hopson had virtually laughed the future president out of the room and that Roosevelt bore a grudge thereafter against the entire industry."

3. In most states, commissioners are appointed by the governor to staggered terms of three to seven years; in twelve states they are elected directly by the public. Typically, commissions are comprised of three to five members; about half require minority party representation. Oregon is the only state to trust responsibility for public utilities to a single commissioner. In New York, the chairman of the state public service commission, who is also the head of the state's department of public service, is designated by the governor; in California, where the commission staff does not function as a part of the executive branch, the commission president is elected by the other commissioners. The rest of the states divide themselves about equally on these organizational arrangements.

   The jurisdiction of most state commissions, like California's, extends to all industries traditionally thought to be public utilities—electric, gas, water, telephone, and telegraph companies—and to some transportation industries such as airlines, bus lines, highway carriers, and of course, railroads. New York and other states have separate commissions to regulate transport industries. The commissions of Virginia and a few others regulate nonutilities like insurance companies, banks, savings and loan associations, credit unions, and brokerage houses. A few commissions even collect motor-fuel taxes, supervise retail franchising, register trademarks, and assess utility property for tax purposes. *See Wall Street Journal*, August 16, 1978, p. 1.

   In the majority of states, utility commissions derive their authority from legislative statutes, but in California, Virginia, and several others, the powers of the commission are enumerated in the state constitution and may only be changed within certain limits by the legislature. In these two states and in eleven others, only the state supreme court may review commission decisions.

   The salaries and size of commission staffs vary widely as well. In 1974 Hawaii's part-time commissioners each received $50 a day for their services, while New York's commissioners were paid $43,050 a year. In 1974 the chairman of the New York Public Service Commission was paid $51,150 a year and received a number of benefits, including a chauffeur-driven limousine. (New York has always paid its commissioners well. In 1907, when its first commissioners were appointed, they were paid $15,000 a year.) The size of commission staffs ranged in 1974 from New Mexico's tiny seventeen-member group to the California Public Utilities Commission's 850-member bureaucracy.

   For these and other details of state utility regulation, *see* National Association of Regulatory Commissioners, *1974 Annual Report on Utility and Carrier Regulation* (Washington, D.C.: National Association of Regulatory Commissioners, 1976).

4. *See* Paul L. Joskow, "A Behavioral Theory of Public Utility Regulation" (Ph.D. diss., Yale University, 1972), pp. 17–81, for a detailed description of these three phases in New York.

5. Ibid., pp. 32–34.

6. Joskow, "Behavioral Theory of Public Utility Regulation," p. 203.

7. For the details of cost-of-service rate-making see Phillips, *Economics of Regulation*, pp. 123–440.

8. Alfred E. Kahn, "Speech given at the New York Society of Security Analysts," New York, December 18, 1974, on file at the New York Public Service Commission, Albany, New York, p. 2.

9. Wilcox and Shepherd, *Public Policies Toward Business*, p. 408.

10. McDonald, *Insull*, pp. 67–69.

11. Wilcox and Shepherd, *Public Policies Toward Business*, pp. 408–413.

12. Ibid.

13. William G. Shepherd and Thomas G. Gies, *Utility Regulation: New Directions in Theory and Practice* (New York: Random House, 1966), pp. 277–278.

14. In the following section, I am indebted to Professor Paul L. Joskow of the Massachusetts Institute of Technology for helpful comments and for his paper "Electric Utility Rate Structures in the United States: Some Recent Developments," presented at the Seventh Michigan Conference on Public Utility Economics in April 1977, and reprinted in Werner Sichel ed., *Public Utility Rate Making in an Energy-Conscious Environment* (Boulder, Colorado: Westview, 1979), pp. 1–22.

15. Kahn, "Speech given at the New York Society of Security Analysts," p. 2.

16. *Business Week*, May 25, 1974, pp. 111–112.

17. In Georgia, Mr. and Mrs. Bernard Tuvlin brought suit against Georgia Power Co. for its promotion of all-electric homes. "The couple, who had a December [1974] electric bill of $211 for their all-electric house, are suing the Southern Co. subsidiary for $1 million in exemplary damages, for $1,759 to reimburse their cost of installing a gas furnace and a gas hot-water heater, and for $25,000 for inconveniences they suffered. Their attorney . . . says that the suit raises the question of whether a public utility may be held legally responsible for its promotional techniques. He says Georgia Power knew that its promotional claim that totally electric homes were efficient was false." *Electrical Week*, February 17, 1975, p. 8.

18. *See* Edward Berlin, Charles J. Cicchetti, and William J. Gillen, *Perspective on Power* (Cambridge, Mass.: Ballinger, 1975), pp. 1–11.

19. National Association of Regulatory Commissioners, *1974 Annual Report*, pp. 548–49.

20. Phyllis Peterson and George C. Lodge, "Consolidated Edison (B)" (Harvard Business School: Case 9-375-130).

21. *See Business Week*, May 25, 1974, p. 102.

22. *See Electrical World*, November 15, 1977, p. 9.

23. *Wall Street Journal*, May 23, 1978, p. 48.

24. The flow-through method is an accounting method under which decreases or increases in state or federal income taxes resulting from the use of liberalized depreciation and the investment tax credit for income tax purposes are carried down to net income in the year in which they are realized. The normalizing method is an accounting method under which decreases or increases in income taxes resulting from accelerated amortization and/or liberalized depreciation deductions in income tax returns are offset in the income account with corresponding credits or charges to balance sheet accounts maintained for accumulating the net balances of deferred and future income taxes. For a more complete definition see Electric Utility Rate Design Study, *Glossary: Electric Utility Ratemaking and Load Management Terms* (Palo Alto Calif.: Edison Electric Institute, 1978), p. 57.

25. The turbulence in the world of regulators which has come as a result of these structural changes is perhaps reflected by the fact that in 1978 there were only

30 regulators nationwide with 10 or more years of experience as compared with 55 such persons ten years earlier. *See Wall Street Journal*, May 23, 1978, p. 48.

26. A generic rate proceeding is one that examines ratemaking principles outside the confines of any utility's specific rate case.

27. Hearings Before the Subcommittee on Energy and Power of the House Committee on Interstate and Foreign Commerce, "Electric Utility Rate Reform and Regulatory Improvement," Ninety-fourth Congress, second session, serial no. 94-127 (Washington, D.C.: Government Printing Office, 1976), p. 1.

28. Economic Regulatory Administration, *Electric Utility Rate Demonstration Program Fact Sheet* (Washington, D.C.: Department of Energy, 1977).

29. Department of Energy, *H.R. 4018 Public Utilities Regulatory Policy Act: Technical Briefing Book* (Washington, D.C.: Department of Energy, 1977). But see also Daniel Hill, Robert M. Groves et al., *Evaluation of the Federal Energy Administration's Load-Management and Rate Design Demonstration Projects* (Lansing: Survey Research Center, Institute for Social Research, University of Michigan, December 1978).

30. H.R. 12461 Ninety-fourth Congress, Second Session, 1976. *See also* Hearings, p. 2.

31. Dingell conducted hearings on his bill in March and April of 1976 as chairman of the Subcommittee on Energy and Power of the House Interstate Commerce Committee. The hearings attracted wide attention in the industry and among its various publics and resulted in a record of over 2,000 pages which served as a basis for portions of the Carter Administration's bill.

32. H.R. 4018 *Public Utilities Regulatory Policy Act: Technical Briefing Book*, p. E-4.

33. Rate structure reform in 1978 was in a fluid state in most regulatory jurisdictions, and it is still too early to say with complete assurance whether policies will converge. In June 1978 the Massachusetts Department of Public Utilities (DPU) ordered time-of-day rates as well as a lifeline-like rate that grants needy customers of Massachusetts Electric over sixty-five years of age a 30 per cent discount on rates. But it is clear that the DPU is only experimenting with these rates. Of the discount rate, commissioner Paul F. Levy said, "The group of people [eligible] is clearly identifiable, small and stable, so we thought we would give it a try. We're not clear we would extend it to other groups." *See Boston Globe*, evening edition, June 1, 1978, p. 25. The Florida Public Service Commission has been so divided over inverted rates that it has alternated on the subject according to the mood of one commissioner. *See Wall Street Journal*, September 19, 1977, p. 8. In 1977 the National Association of Regulatory Utility Commissioners reported that 30 state commissions were actively considering new types of rate structures. *See Wall Street Journal*, September 19, 1977, p. 8.

Chapter 4

THE POLITICS OF EFFICIENCY: ADOPTING MARGINAL-COST-BASED ELECTRIC RATES IN NEW YORK STATE

The only economic function of price is to influence behavior. This is a notion that traditional regulators have great difficulty accepting.

ALFRED E. KAHN

On January 29, 1975, the New York State Public Service Commission (NYSPSC) issued orders instituting a "generic" rate investigation "to inquire into the merits of, and to develop principles and methodology for, the revision of electric rate schedules."[1] In the preamble to its order, the commission stated, "Rapidly increasing costs of new generating facilities and the rising cost of fuel make it urgent, in the interest of energy conservation and the efficient use of resources, that the structure of energy prices reflect, to the greatest extent feasible, the variations in the incremental costs of service because of differences in the time of consumption, as well as in all other cost-influencing factors."[2]

A little over a year and a half later, after reviewing the record compiled in 35 days of hearings and the briefs of 22 parties, the members of the commission unanimously concluded that "marginal costs . . . provide a reasonable basis for electric rate structures" and ordered each of the electric utilities within the commission's jurisdiction to develop "studies sufficient to translate marginal cost analysis into rates."[3] Within six months the commission had examined and found reasonable the state's first mandatory time-of-day rate based on marginal cost. The rate was offered by the Long Island Lighting Company (LILCO) and was to apply to the utility's 175 largest commercial and industrial customers. The rate became effective February 1, 1977.[4]

The immediate impetus for the generic hearing came from separate petitions filed by two unlikely co-sponsors: the Environmental Defense Fund and the state's seven major investor-owned electric utilities. The EDF wanted to use the generic hearing to establish New York as the leading state in a nationwide effort to slow down the growth in energy consumption by reforming electricity rates. The utilities wanted to avoid dealing with questions of restructuring these rates in the context of their individual applications for permission to increase rates. They felt that the principles of rate design should be developed in a general investigation. Some, but not all of the utilities, believed that new methods of pricing might enable them to recover their costs in a more efficient manner.

The principal force behind this investigation and the commission's decision to embrace marginal cost, however, was Alfred E. Kahn. Kahn was appointed chairman of the NYSPSC in July 1974; he would later serve under President Carter as chairman of both the Civil Aeronautics Board and the Council on Wage and Price Stability. He had been professor of economics at Cornell and the dean of the university's College of Arts and Sciences. From the day he was made chairman, Kahn had a clear understanding of the principles that would guide his regulatory efforts. He had spent more than twenty-five years teaching such courses as "Private Enterprise and Public Policy" and in 1970 had published *The Economics of Regulation: Principles and Institutions*, a highly regarded, two-volume treatise.[5] For Alfred Kahn, the value of marginalist principles for guiding pricing decisions was unquestioned. However, he arrived on the scene with an open mind as to the feasibility of marginal cost pricing in regulated industries and with a recognition that certain notions might not be politically or administratively attainable.

It is tempting to conclude that Kahn's well-known beliefs alone were sufficient to insure the success of marginal-cost-based rates in New York. The problem with such a conclusion is that it fails to explain the notoriously poor showing which typically characterizes the efforts of economists to translate economic reasoning into public policy. If we accepted this view we would fail to ask whether or not, and why, Kahn would have been as successful in 1964 when fellow economists were despairing of ever changing the declining block structure as he was a decade later—two questions that must be asked. Moreover, contenting ourselves with this conclusion reinforces the common misunderstanding that all one needs to know to explain political behavior is some information about the motivation or goals of political actors. Constraints on the behavior of governmental executives, of the sort Kahn anticipated when he accepted the post of NYSPSC chairman, are often overlooked by analysts. As was argued in Chapter 1, these constraints may well be the determining factors of regulatory policy.

The efforts of the NYSPSC under Kahn to restructure electric rates in accordance with the requirements of economic efficiency provide us with an opportunity for evaluating the determinants of bureaucratic regulatory behavior when internal constraints are likely to take on increased importance because the task at hand is of the "loosely constrained, planning" variety (Type IV in the typology of Chapter 1). The construction of a marginal-cost-based rate structure is not something a commission can decide to institute today and then effectively ignore tomorrow. Numerous questions about the methodology for determining marginal costs, the planning period to employ, the effect on system load factors, and the response of customers to new prices continually involve the commission in coordinating the activities of staff members and utility rate experts and assessing the usefulness of information contributed by outside intervenors. Marginal costs, unlike historic or average costs, have to do with both present and future; they cannot be ascertained simply by auditing a utility company's books. Future costs must be appraised and reflected in current rates so that customers can judge whether the value of the additional electricity they intend to consume is worth the cost to society of producing it. These costs are necessarily hypothetical and subject to change, requiring a good deal of judgment on the part of the commission's rate experts.

Given the complexity of the task and its radical departure from traditional ratemaking procedures, we might ask the following questions: Why was Kahn successful at all? What precisely were the bureaucratic and political constraints Kahn faced? How were they overcome? To what extent was Kahn able to achieve his obvious goal of bringing electricity rates in line with marginal costs?

The New York Regulatory Environment

We can gain some insight into the constraints and opportunities that Chairman Kahn faced in 1974 by evaluating the New York regulatory environment that predated his appointment as chairman of the public service commission. During the sixties the New York State Public Service Commission, like most utility commissions, was a sleepy outpost for political patronage. As one influential commission source observed:

> *I think that if you look closely at the composition of the commissions in the 'sixties and the 'fifties—which I have—it's not too difficult to see that they weren't very good. The commissioners were almost all political appointees—either defeated candidates or retiring politicians who were basically in need of a job. And it's evident in the performance of the companies. First of all there were very few rate cases—with Con Ed being about the only exception—and they were small. . . . Companies were earning well in excess of [a reasonable] amount and there seems to be evidence that they were expanding capacity at a greater than optimal rate.*

The commission and the utilities were able to "get away" with poor performance because the real price of electricity was either stable or falling:

> *Consumers didn't realize they were paying high rates of return to the companies—or care much because their utility bills were small. . . . I've concluded that in good times the governor can get away with making patronage appointments to commissions like this. The governor always has a certain number of patronage appointments, and a good governor will look around to find the place where he can make these appointments without doing too much damage. I think from that standpoint the PSC was a good place for patronage appointments in the sixties and the fifties.*

In New York this lazy ambience was rudely interrupted in 1969 by a series of rate increases, service failures, and legislative investigations:

> *Con Ed came in with a major rate case in 1969 and New York City began experiencing some brownouts. Then there were some real telephone service problems in New York City in 1969—all of which led to greater public attention to the PSC. The legislature [that year] began holding hearings on the PSC and in the course of the hearings it became clear that the then-chairman of the PSC didn't really know what was going on in his own department and that proved to be a real embarrassment for the Governor.*

The Appointment of Joseph C. Swidler as NYSPSC Chairman

In response to the shortcomings of the NYSPSC, Governor Nelson Rockefeller took steps in early 1970 to strengthen the position of chairman by finding a national figure to fill the office. His choice was Joseph C. Swidler, the respected former chairman of the Federal Power Commission (FPC). Swidler had served the FPC from 1961 to 1966 and before that had been general counsel to the Tennessee Valley Authority. While at the FPC, Swidler had instituted the National Power Survey and consequently was strongly identified with electric reliability. Late in 1969 he had again been in the news, having participated in a large study of the power requirements of the Northeast. Swidler accepted Rockefeller's invitation to rebuild the NYSPSC and was appointed February 1, 1970.

Swidler soon became a commanding figure at the commission and at the Department of Public Utilities, which was reorganized to come under the administrative direction of the NYSPSC chairman (see Figure 4.1). Almost immediately after he arrived, Swidler made a number of organizational and programmatic changes. Systems planning was introduced for the first time in the commission's communications, power, and gas divisions. The former head of the Department of Public Utilities, a man staff members described as very authoritarian, was eased into retirement and the interchange between staff and commissioners greatly opened up at regular weekly meetings. A number of Swidler's best assistants from the Federal Power Commission followed him to New York and were appointed to key positions on the commission staff. An innovative new manager in the communications division completely eliminated the service problem that had drawn so much negative attention to the commission in 1969, and in his first year at the commission Swidler encouraged the staff to begin presenting an affirmative case in rate hearings for the first time.

More important in terms of Swidler's influence on Kahn's later attempts to restructure rates in accordance with marginal cost was Swidler's authorization of an extensive "fully allocated cost study" by the staff which was to provide the commission with an independent basis upon which to allocate responsibility for a utility's revenue requirement by customer class. This was the first step the commission took in the direction of designing electric rate structures. Innovative as it was, the fully allocated cost study was not based on

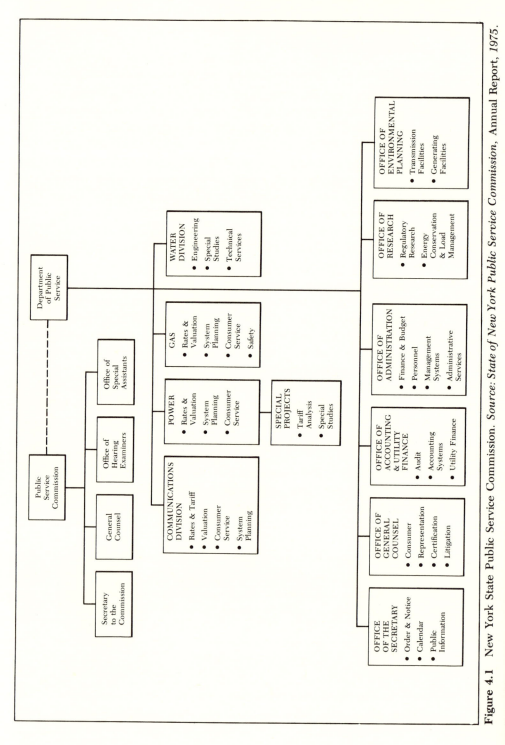

Figure 4.1 New York State Public Service Commission. *Source: State of New York Public Service Commission, Annual Report, 1975.*

principles of marginal cost. Indeed, Swidler was an outspoken critic of marginal-cost-based rates after leaving the commission, as evidenced by a 1975 interchange between Swidler and Irwin Stelzer, president of National Economic Research Associates (NERA). [NERA is an organization of consulting economists that has delivered testimony in many state generic hearings, including those in New York, in support of long-run incremental cost (LRIC), a marginal cost approach to electricity pricing.] In response to a presentation by Stelzer at a rate conference sponsored by the Public Utility Research Center of the University of Florida, Swidler rose to say that LRIC was based on a "doubtful allocation" of costs, an allocation "compounded by [an] illusion of preciseness." He argued that LRIC could cause "problems of dislocation [in the economy] that could be very severe" and warned that its practical application involved value judgments and many trade-offs "which could be for the worse." Finally, he said he spoke against LRIC because it would put regulatory commissions out of work "and substitute instead a bunch of fuzzy-cheeked economists with their econometric models" to determine how costs should be allocated and how high electric bills should be.[6]

Swidler professionalized the New York staff between his arrival in February 1970 and his resignation in May 1974, instilling an esprit de corps that caused many staff members to consider themselves to be working for the "premier utility commission in the country." Certainly, Swidler influenced the attitudes of many of the commission's engineers on the matter of cost-of-service ratemaking—a matter that created both problems and opportunities for his successor, Alfred Kahn.

Consolidated Edison and the OPEC Embargo of 1973

In late 1973, Swidler announced in a memorandum to the commission staff that he would be leaving the commission soon. One of Swidler's colleagues on the NYSPSC, a person who had been with him in Washington, explained his decision to resign in the following way:

> There were a combination of factors [which led to his decision]. In late '73 and early '74 Rocky was being talked about for various high level jobs in Washington, so that his commitment to the governor was no longer as active as it had been. Then the oil embargo hit in October

1973, creating enormous problems for the utilities. It would have taken a great deal of energy to ride that through, and I think he was basically tired. . . . The pace was hectic. At the time he was 67.

Although an interdepartmental task force comprised of representatives of New York's environment, commerce, and transportation departments had been set up in 1972, with the NYSPSC chairman as its head, to evaluate the impact of an energy shortage, the state and the electric utilities that operated within it were caught unprepared by OPEC. Consolidated Edison, which supplies electric power to New York City's five boroughs and Westchester County, was especially hard hit. The company's problems had in part been instrumental in bringing Swidler to the commission; now they created enough headaches to encourage him to leave. When Con Ed decided to pass on its dividend in the spring of 1974, it lost the confidence of investors. When company chairman Charles F. Luce recommended selling two unfinished nuclear plants to the Power Authority of the State of New York (PASNY), it lost the confidence of industry executives. Donald C. Cook, the chairman of the American Electric Power Company, went so far as to say: "Con Ed is out of the equity market and New York State is in the power business—forever." Other executives even expressed suspicion that Luce, who had once headed the federal government's Bonneville Power Administration, and Swidler, who had worked at the TVA for 20 years, were biased against private power and were using the crisis as a cover for the take-over offer: "Luce has always felt that public power is better than private power. And the chairman of the state commission is a public power man. What they seem to want is a TVA system in New York."[7]

But if the utilities thought that Joseph Swidler was going to use the energy crisis to expand public power in New York, they were wrong. In little more than a month after Con Ed passed on its dividend, Swidler was out, and Alfred Kahn was in.

The Victory of Marginal Cost Pricing

When Alfred Kahn took over the Department of Public Service as Joseph Swidler's replacement in July 1974, he inherited a hard-working, professional staff, but they were people who had to be convinced of the value of marginal-cost-based rates. He did have some allies, however, and the timing seemed to be right. The Environmental Defense Fund had already intervened in a Wisconsin rate case on behalf of time-of-day rates, and so was beginning to put together a credible argument. The utilities, while deeply suspicious about any regulatory intervention in their electricity marketing practices, were nonetheless strapped for cash and looking for a way to conserve on their capital requirements. Large users were perhaps most unitedly hostile to the reform of rate structures. Having enjoyed the benefits of promotional rates and big discounts for years, they viewed with animosity every effort that suggested an end to their special privilege. Finally, the state's political leadership was in a state of flux. With Rockefeller leaving for Washington to serve as Vice-President it was too early to tell what might be placed on the public agenda after the November elections. Reforming utility rate structures was a formidable undertaking in 1974, but due to changes in the regulatory environment a "politics of efficiency" was at last feasible. To make its success probable, Kahn had to have the support of the staff.

Winning the Support of the Staff

It would be false to create the impression that Alfred Kahn approached the task of adopting marginal-cost-based rate structures with a complete understanding of the political pitfalls and opportunities before him. Nevertheless, whether by design or by habit, Kahn did seek first to convince his colleagues on the staff that rates should be based on marginal, not average, cost. While he did not convince everyone—most notably, the rate engineers in the gas division, where a Swidler appointee was especially influential—he did find enough support among the engineers in the electric power division to help him advance his concept.[8] He won their respect and even their enthusiasm with his thoroughgoing professional manner; with his constant appeals to their own professional norms that rates should be based on costs; with his offer of an apparent solution to one of their most vexatious problems, that of "revenue erosion"; and with his nonthreatening, familiar, and congenial personal style.

It was Kahn's style that staff members were apt to cite first in interviews given after Kahn had left the commission in 1977 to become chairman of the Civil Aeronautics Board. One staff member reminisced:

The primary reason Fred [was able to convince us] was that he was so darn enthusiastic about it and he knew exactly what he wanted. Plus, he worked awfully hard and was a bit flaky—which helped. In fact, we were just talking about what we should wear to the dinner that the staff is having to honor Fred. We decided that if you had a coat which had one sleeve a little loose, a tie that didn't quite go with it, and a pair of pants that didn't match at all—which were preferably crumpled up—that that would be just about right. . . .

He was unpretentious and at ease with people. He didn't tell you what to do; he often didn't ask—he'd suggest. He was on a first name basis with a lot of people around here. There was always a lot of give and take in commission sessions between the staff and the chairman. . . . He was exceptionally quick—having run all those graduate seminars. . . . People respected him.

Many staff members found Kahn's penchant for "running seminars"—as he called them—refreshing and enlightening. A staff member noted, "Fred wasn't at all concerned about getting out the chalk board and explaining his theories to the commission during meetings if he needed to."

Even Kahn's generic investigation on rate design departed from the traditional format of rate cases. He organized witnesses in panels, with each panel composed of individuals with divergent views. All of the attorneys sat facing the panel and took turns asking questions seriatim, after the witnesses had offered their testimony. Persons were encouraged to break in and ask clarifying questions whenever they wished to do so.

Not all of the attorneys appreciated this style of the investigation, and those who represented industrial or commercial customers found it particularly objectionable. At one point a group of industrial consumers who objected to Kahn's handling of the generic hearing filed a formal motion with the commission to recuse Kahn for his "preconceived notions regarding the appropriate elements of rate design for electric service"; for having "departed from his quasi-judicial role as a regulator"; for having "indulged in extensive cross-examination," having made "extensive extrajudicial comments on the record," and having "clearly indicated impatience with views [he] considers unacceptable"; and for having showed a "propensity to inject testimony."[9]

The commission denied the industrial consumers' motion. In his own defense, Kahn explained the purpose of the departure from the traditional format:[10]

I have adjured all parties to look upon at least the theoretical phase of the proceedings as an intellectual exercise, to treat it in the nature of a running seminar, in which our purpose was essentially to explore certain academic ideas. That is the spirit in which I have behaved whenever I have presided. I have never pretended to play the role of a passive receiver of evidence. I made clear from the beginning my intention to participate actively in the process of exploration. It was for this reason that I joined in ruling that witnesses be grouped in panels . . . so that they might evaluate one another's ideas, respond to one another and to me, to provide us with different opinions and ways of looking at questions as they arose. I intervened freely—attempting, however, not to interrupt the train of a lawyer's cross-examination—to make certain that I understood what a witness was saying and what he was not saying, to test his ideas, to offer hypotheses of my own in order to get his reaction. In all cases, I invited all members of the panels to partake freely, to comment on my comments and on each others'. I urged them to volunteer reactions, promising to recognize them at any point, consistent merely with orderly progress. I propounded problems—questions of whose answers I was myself uncertain; and when the witness was uncertain about the answer I asked him, and all the others, to think about it. No witness, of whatever point of view, was ever denied an opportunity to express his opinion, as fully as he wished.

If his style prompted legal challenges from various intervenor groups, it won Kahn support on the staff. An associate valuation engineer in the power division offered this judgment:

Under Kahn for the last three years it's been great. First of all he's a totally honest individual. He's not interested or affected by any of that political stuff. He's run the commission in a very forthright, honest way. I know Fred Kahn. Even an individual at my level is on a first name basis with him. He's personable, but more important he's generally concerned about doing the right thing and he's given us the space to try to do it. It's been a privilege having him as chairman.

Unlike Swidler, Kahn made few major organizational or personnel changes. Instead, he identified a few key people who were receptive to his ideas and worked through them. This strategy helped to assuage fears and gain friends on the staff, but it, too, was criticized from outside the commission. From the governor's office came this criticism:

Fred Kahn is a fine, nice, personable guy—but not a terribly good manager. He's just too nice. He never in the entire time he was there fired anyone or demoted anyone, including some who should have been.

He would bring people into his office and if they didn't know anything about marginal cost he'd sit them down and patiently explain it to them as if they were a freshman—congenial, never exasperated, always patient. Of course, if somebody was quicker, then he'd pick up from there, but always a gentleman. He'd never blow up at anyone and say this is unsatisfactory or this won't do, etc. Everybody loved him and that was one of the problems. He could never get rid of some of the dead wood—and I think a good manager would have done that.

The Triumph of Professional Norms

Kahn's personable style certainly helped smooth the way for the acceptance of his ideas on marginal costs, but his style alone was not sufficient to convince the staff members—many of whom were engineers and found Kahn's views to be a new and alien way of thinking about utility regulation. Many staff members initially shared Swidler's views that marginal-cost-based rates were judgmental and uncertain. While the engineers defined their job as achieving efficiency, they took "efficiency" to mean setting electric rates to encourage the most technically efficient production of electricity (unlike economists who take it to mean the best allocation of society's resources). This, in turn, meant that the engineers typically favored volume discounts for large users, either to achieve economies of scale or to avoid excess capacity by discouraging industrial and commercial users from shifting to generating their own electricity.

Eventually, the engineers became convinced that rates should vary according to time of use, but they were less than fully convinced that time-of-day rates should be based on incremental or marginal costs rather than on the "embedded" costs traditionally used to calculate the revenue requirement. Some engineers felt that although rates based on marginal costs require peak load pricing, the converse was not true: It was possible, some maintained, to justify and base peak load rates on average costs as well as on marginal costs. The chief advantage of time-related rates based on average costs from the point of view of utility engineers was that the costs could be verified by examining a company's books. Moreover, the average cost methodology corresponds more closely to the setting of the revenue requirement—a task that utility engineers know how to perform.

"You've got to remember that this is new to engineers," said one of the non-engineers on the staff. "All those years of human capital

investment are being threatened, so you can expect them to resist."
"The engineers weren't used to doing things Fred's way," said one of
Kahn's fellow commissioners. "They were being asked to think like
economists, which they had no trouble doing; it's just that that
wasn't their training or practice."

But engineers have a strong professional norm to base rates on
costs. Kahn both understood and shared that norm in making his
pitch for marginal cost. "Basically he appealed to their sense of
professionalism—he was clearly very professional in his own
approach—through the simple principle of encouraging them that
what we wanted to do was get prices to conform to and track costs—
that simple principle was very appealing."

Robert Smiley, an economist from Cornell whom Alfred Kahn had
recruited as his special assistant and who was later director of the
commission's office of research, elaborated:

> *One of the very appealing things Fred used effectively over and over
> again in dealing with engineers who had developed this ratemaking
> formula to precision was the question: "Do you want to be precisely
> wrong or approximately right?" One of the problems, of course, was
> that they spoke a different language from us. But he was successful in
> the power division, I think, because of the power of his intellect and
> his continual appeal to the notion of getting price in line with costs.
> Then, too, he had a clear sense of what it was he wanted, and he
> devoted a lot of attention to it.*

In the end, however, the engineers were won over to Kahn's
position because they saw in Kahn's principles a way to solve one of
their most troubling problems, which they termed "revenue erosion." The term refers to the erosion of the financial position of
utilities which occurs when growth in expenditures exceeds growth
in revenues. Revenue erosion is precisely the problem in cases in
which electricity is priced at less than marginal cost. Although they
did not understand the problem in these terms, staff engineers
did realize that revenue erosion was due to excessive demand
growth in the tail blocks of utilities' declining block rate structures.
In 1972 a staff memorandum blamed revenue erosion for the increased number of rate cases the commission was asked to decide:[11]

> *When increased sales do not provide appropriate increased revenue, it
> is evident that . . . tariff structures have not been properly designed
> and, unless there are offsetting decreases in expenses, utilities will fail
> to earn allowed rates of return. . . .*

The classical "across-the-board" approach [of increasing rates], that is, applying the same percentage increase in each rate block, adds to these problems . . . [and] may be a major reason why revenue erosion continues after substantial increases in rates, requiring utilities to seek further rate relief, sometimes almost immediately. . . .

These methods must be abandoned in favor of adjusting each rate block after a thorough analysis of the . . . cost of providing service in the individual rate block.

In mid-1974 the commission staff was desperate to find a way to hold down the number and size of rate increases. The engineers began to perceive marginal-cost-based rates as a way to deal with revenue erosion and reduce the magnitude and frequency of general rate increases. Jack Treiber, one of the engineers who became convinced of the importance of marginal cost analysis, made the following summary observation:

Fred started holding meetings on rate design for anyone interested. I was doing rate design work, so I went. He was the professor and we were the students. We didn't immediately go with it. "What is this marginal cost?" we asked. Finally, it became clear that this was just what we needed to deal with the revenue erosion problem. We were basically there—we just didn't have the buzz words. We needed an economist to give us the buzz words.

Fred's personality, his professionalism, certainly moved things along. But it would have come anyway. What he gave us was a set of principles to deal with a problem we were having. It just seemed so logical—the way to get from A to B. One could easily argue in favor of marginal cost. It was less easy to argue against—you'd have to bring up side issues that weren't that important.

Issues Arising From Marginal-Cost-Based Rates

Kahn realized that winning the commission staff's enthusiasm for the principle of marginal cost, while necessary, was perhaps the least difficult part of the battle before him. Almost prophetically, in 1970 he had written:[12]

These are the principles of economically efficient public utility pricing. Formulating them was the easy part of the job. Most public utility executives and regulators would probably acknowledge their validity, while taking pains at the same time to point out that economic efficiency is not the sole test or purpose of their performance. But they would also hasten to emphasize that these principles fall far short of providing workable rules for the guidance of their accountants or engineers. The task of translating these principles into actual schedules is so extraordinarily difficult that it is entirely possible to accept their validity while at the same time concluding that the task of following them is an impossible one. Few would go as far as to abandon the effort entirely. But all would point out, and correctly so, that even the most sophisticated and conscientious effort to apply these principles inevitably involves large doses of subjective judgment and, at the very best, can achieve only the roughest possible approximation of the desired results.

A number of difficult issues needed to be addressed in the generic hearing and in specific cases before a new rate structure based upon marginal costs could be designed. Some of these issues were primarily technical; others had important political implications. Chief among these issues were the following: (1) the question of whom to meter for time-of-day usage; (2) the problem of "second best"; (3) the problem of the "shifting peak"; (4) the related problem of "needle peaks"; (5) the definition of marginal costs; (6) lifeline rate proposals; (7) the disposition of excess revenues; and (8) the electric industry's charge of discrimination and its threat to relocate in response to new rates.

Whom to Meter?   In a 1976 report on the progress of utility regulation in New York delivered before the American Economic Association, Alfred Kahn observed:[13]

The only economic function of price is to influence behavior. This is a notion that traditional regulators have great difficulty accepting. . . . But of course price can have this effect on the buyer's side only if buyer's bills do indeed vary depending upon the amount of their purchases. For this reason, economists in [the] public utility field are avid meterers; and if they were costless, we would like to have meters just as complicated as necessary to measure every dimension of consumption that has an independent marginal cost—numbers of telephone messages, the time each one is placed, the number of minutes each one consumes, the number of miles it traverses, the number of kilowatts and the instant at which each is taken, the number of feet of distribution system required to serve each customer; and so on.

Kahn went on to observe, however, that economic efficiency requires that the "incremental costs of progressively more complicated meters required to administer progressively more efficient pricing schedules [be compared] with the incremental benefits of those refinements."[14]

Because time-differentiated rates would require new meters for all customers to whom the new rates would apply, one of the issues the commission needed to address was whom to meter. For large customers, even rough cost-benefit analysis suggested the advantage of metering, but for small customers the answer was less clear. In 1978 new time-of-day residential meters cost nearly $150 whereas standard meters cost only $23.[15]

The Problem of "Second Best." A familiar theoretical objection to applying marginal cost principles in designing rate structures is the problem of "second best." In brief, anti-marginalists argue that in a world in which other prices deviate from marginal costs it is impossible to conclude as a general proposition that setting prices equal to marginal costs in any single sector will be desirable. Anti-marginalists maintain that without knowledge of the extent to which all other prices in the economy besides that of electricity deviate from marginal costs there can be no certainty that the use of marginal costs in designing electric rates alone will improve the allocation of resources since it is the *relative* prices of various goods and services which are pertinent in influencing the behavior of buyers.[16] Facing the commission, then, were the two related questions: "Is the problem of second best sufficiently important to render any change in rate structures harmful? If not, what deviations from marginal cost should be made in light of second-best considerations?"

The Shifting Peak. If electricity taken on-peak is priced higher than that which is taken off-peak, a possible customer response is to defer some consumption to off-peak periods. Indeed, for environmentalists and others, this response is the chief advantage of peak load pricing inasmuch as it improves load factors and diminishes the need for additional plant capacity. One practical problem, however, is that customers may shift so much of their consumption off-peak that a new peak is created, necessitating a further revision in the rate structures.

The issue of the shifting peak had been recognized by other utility commissions. Wisconsin public service commissioner Arthur L. Padrutt, dissenting from the commission's majority, had decided that peak load pricing, "In theory . . . is excellent, but it should be very obvious that, if in practice the gambit is successful, the ultimate result wipes out the 7 P.M. peak and builds a new one at 10 P.M. . . . A most expensive investment in metering equipment would add a substantial sum to customer costs for the dubious benefit of chasing peaks around the clock."[17]

Needle Peaks. If seasonal and time-differentiated rates succeed in discouraging some peak consumption, they may still create "needle peaks" that could destroy any improvement in utility load factors. A possible needle-peak-creating scenario is the following:

> Con Ed institutes a seasonal and time-related rate for its New York City and Westchester County customers. The utility has a summer peak, so rates from noon until 6:00 P.M. in July and August are especially high. The high rates discourage customers from using their air conditioners on all but the three or four hottest days of the summer, but on those days they are turned on full force, creating a huge surge in demand. Since Con Ed has no way of knowing in advance precisely which days in the summer will be the hottest, its rates cannot fully discriminate between those who are responsible for the needle peak and those who take power on other summer days.
>
> The result of this lack of discrimination is that rates on the hottest days are too low to cover the incremental cost of service, thereby encouraging excessive demand. Meanwhile, rates on other summer days are too high, which discourages some consumption that would improve the utility's load factor.

An optimal rate structure would differentiate by temperature at time of use. Even if this differentiation did not improve load factors, it would at least assign the burden of the marginal costs to those who were responsible for them. In the absence of a temperature-sensitive rate, the commission needed an alternative strategy for dealing with needle peak problems.

The Definition of Marginal Costs. Defining marginal cost in an effort to price electricity can be both complex and confusing. The familiar proposition from elementary economics that "sunk costs are sunk"—and therefore irrelevant to incremental price and output decisions—focuses attention on the variable costs of production in plants already in existence. Short-run marginal cost is the addition to total variable cost associated with the production of the next incremental unit of output. The customers who should be charged the marginal cost are those who are responsible for the increase in variable costs. If they are charged the full marginal cost, such customers can determine for themselves whether the extra cost they impose on society is worth the benefit they derive from the last unit of output they consume.

These principles are simple in the abstract, and they give us a clear pricing rule: Prices should be set to equal short-run marginal costs on the smallest increment of output possible. However, the principle of causal responsibility implies that price must include all of the variable costs, regardless of when they are realized. If additional production today will cause capital equipment to be utilized more intensively and therefore to wear out faster, the costs of replacing that capacity should be included in the variable costs. For this reason, in practical application marginal cost pricing may consist of setting rates equal to long-run incremental cost (LRIC). LRIC measures the change in total costs when output is increased or decreased by an increment of a block of output for an extended period of time during which system capacity can be altered. How long this period should be—that is to say, how many costs should be considered variable—will depend upon the planning perspective utilities and regulators employ, something that obviously will vary from one instance to another and may be a matter of dispute.[18]

Consider, for example, the question of charging current peak users for a nuclear plant that may not begin service for ten years. Should senior citizens who take power on-peak but who will very likely be dead in ten years be required to pay for the plant in their rates? The economic principle that could serve as a guide in this case is a simple one: Charge the customer who is causally responsible for the new plant being built. The determination of causal responsibility, however, depends on the planning period selected and is therefore a matter of judgment.[19]

A related example pertains to the assignment of responsibility for the cost of additional plant capacity. In the short run, when plant capacity is considered fixed, marginal cost pricing would ignore capacity charges; but if it is determined that a longer planning period is justified, a method must be devised to assign these costs among various users. The economic principle to be applied again is clear: If the same capacity serves all users, capacity cost should be levied only on those whose use requires the expansion of capacity, the peak users.[20]

In the production of electricity, however, the same capacity does not necessarily serve all customers. Generally, utilities cope with diurnal and seasonal variations in the demand for electricity by employing several types of generators, some of which are constantly in use—baseload plants—and some of which are used for shorter periods of time—cycling and peaking units. Companies do this to minimize the total cost of providing service at a specified level of reliability. Baseload generators (nuclear and coal generators) have high capital costs but relatively low operating costs, whereas peaking units (gas-fired turbines) have low capital costs and high operating

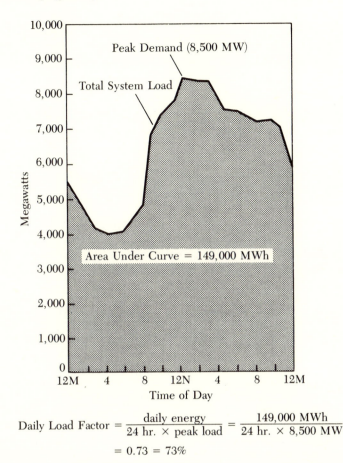

$$\text{Daily Load Factor} = \frac{\text{daily energy}}{24 \text{ hr.} \times \text{peak load}} = \frac{149,000 \text{ MWh}}{24 \text{ hr.} \times 8,500 \text{ MW}}$$

$$= 0.73 = 73\%$$

Figure 4.2 Daily Load Curve. *Source: Robert G. Uhler,* Rate Design and Load Control *(Palo Alto: Electric Utility Rate Design Study, November 1977), p. 15.*

costs. A typical daily load curve is illustrated in Figure 4.2; the generation dispatched to meet a cyclical load is shown in Figure 4.3.

Since utilities construct gas turbines to meet a small increment in demand, it is the capacity charges for the turbines which should be passed on to peak users. Suppose, however, that a utility has a suboptimal mix of generating plants—that is, that it has too much peaking or cycling capacity and not enough of the more efficient baseload capacity. Marginal costs comprehend only current and future costs. Should an optimal rate structure include any capacity costs under conditions of excess peaking capacity but insufficient baseload capacity?

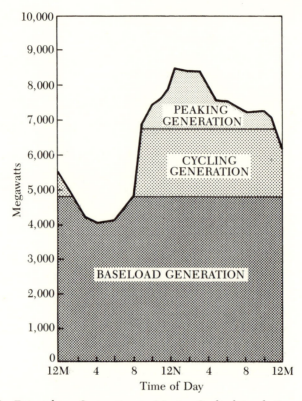

Figure 4.3 Dispatching Generation to Meet a Cyclical Load. *Source: Robert G. Uhler,* Rate Design and Load Control *(Palo Alto: Electric Utility Rate Design Study, November 1977), p. 15.*

The example cited above of the utility with too little baseload capacity is not merely hypothetical. As a result of the enormous increases in the price of oil which followed the OPEC embargo of 1973, New York's downstate utilities felt they had too much oil-fired capacity and not enough coal or nuclear capacity to minimize total costs. Their short-term construction plans consisted only of substituting baseload capacity; no net additions to plant capacity to meet peak requirements were planned.

If the issue of defining marginal costs is troubling for regulators, it is at least a familiar issue. Any ratesetting rule encounters the problem of whether costs may be recovered if incurred inefficiently.

Lifeline Rate Proposals.   Lifeline rates have been described as having "a low-cost initial block for residential users on grounds of 'need.' Everyone, the argument goes, needs a certain amount of electricity . . . [and] proponents of the lifeline concept contend that society should provide for such needs through differential pricing, imposing larger burdens on other presumably more affluent users of electricity."[21] Such rates are clearly not intended to serve the goal of economic efficiency, but electric rates have never been designed exclusively to foster the goal of efficiency. The issue before the commission was that of determining the extent to which rates based on marginal cost should be modified to include other social goals, such as those advocated by the lifeline proponents.

The Disposition of Excess Revenues. Rate structures based on marginal costs are designed independent of the process that sets a utility's overall revenue requirement. Only by chance will marginal-cost-based rates return to the utility precisely the amount of revenue deemed reasonable for it to earn. It is much more likely in an inflationary world that marginal-cost-based rates will generate revenues in excess of those the utility is allowed to earn by traditional ratemaking rules. (Regulators in most states value a utility's rate base at "original cost"—the price paid by the utility to purchase the assets in the first place—or at their "fair value"—a cost typically higher than original cost but less than the cost of replacing the assets.) Thus the question arises, What is to be done with the excess revenues if the utility is not allowed to keep them? Some means must be found to reduce electric rates without distorting the price signals that are the raison d'etre of marginal cost.

The economic principle that could guide the disposition of such revenues is the so-called "Ramsey rule" (or "inverse elasticity rule"), which suggests simply that the extra revenues should be rebated to consumers in inverse relationship to their elasticity of demand. The less responsive to price changes customers are, the greater their reduction in rates should be, and vice versa.[22]

While the inverse elasticity rule is, again, a simple and noncontroversial proposition in the abstract, its application is both difficult and controversial. First, the rule requires a large amount of information about customer responsiveness to price changes. While much empirical research recently has been directed to this issue, the issue has not been settled.[23] Second, despite the primitive quality of demand elasticity estimates, industrial and commercial users fear that regulators will operate on a hunch that the demand of residential consumers is more inelastic than their own and use the inverse elasticity rule to shift to them greater responsibility for utility revenue requirements. Lifeline proponents are quick to make this assessment as well, and so have been wholehearted advocates of the inverse elasticity concept. They argue that their proposals are compatible with economically efficient rates—even though economic efficiency is not their primary goal—because residential demand for the first block of electricity is almost undoubtedly most inelastic.[24]

Industry's Charges of Discrimination and Threats to Relocate. Finally, the application of any new methodology to the setting of electric rates is certain to be criticized as discriminatory if it is not to be applied universally. The commission needed to address the issue of discrimination and also had to face industry's threat that it—and thus jobs—would leave the state if electric rates proved to be too unfavorable.

Most of these eight issues were addressed formally in the generic rate proceeding the commission ordered at the end of January 1975. The record of that hearing could easily be mistaken for that of a somewhat formalized postgraduate seminar. But it is the informal regulatory process that tells us even more about why issues were decided as they were in New York.

The generic proceeding itself had an informal dimension. Shortly before the commission ordered the proceeding to begin, both the Environmental Defense Fund (EDF) and the state's seven largest privately owned utilities petitioned the commission—within a week of each other—to undertake a general investigation of various proposals for reforming rate structures. The motivations of these groups merit description.

The Environmental Defense Fund

The Environmental Defense Fund's interest in rate reform as an alternative to electric power industry growth dated from its intervention in 1973 in a Wisconsin case. Five months before the New York commission ordered the generic hearing, the Wisconsin Public Service Commission, acting largely on a record developed by the EDF, became the first commission to adopt the principles of long-run incremental cost as a basis for designing electric rate structures.[25] New York was the EDF's next target after Wisconsin. The organization had tried intervening in a Niagara-Mohawk case in early 1974 with only partial success. Kahn's subsequent appointment as chairman encouraged them to seek a generic hearing.

Edward Berlin was the EDF's counsel in the Niagara-Mohawk and the Wisconsin rate cases. In March 1975, New York Governor Hugh Carey appointed Berlin a member of the New York Public Service Commission. After he was confirmed, Berlin took no role in the generic hearing before the commission at the time, considering such involvement improper in view of his public advocacy of basing rates on marginal cost principles. A year earlier Berlin had co-authored a book, *Perspective on Power*, with Charles J. Cicchetti (the EDF's chief economic witness in electric rate structure cases) and William J. Gillen in which he presented a framework for the reform of electric utility regulation.[26] Berlin explained EDF's early strategy on rate reform as follows:

> *I had been the counsel for EDF for a number of years—really from the early part of the organization's life when they were just getting started [in the late sixties]. They, of course, had their own counsel, but my law firm represented them in litigation. At that time it wasn't very big—it was mostly a voluntary organization with one or two lawyers and some Ph.D.'s. EDF was involved in a number of cases having to do with energy supply. They were active in the Four Corners case, and I was also involved with them in the Calvert Cliffs case.*
>
> *But it seemed to me that we were just banging our heads against the wall. We weren't having much success, because even if we could make a case that a given facility wasn't needed at the time, it appeared that it was almost for certain that it would be needed in the near future. As a result of that I tried to come up with some alternative strategy. . . . I was very concerned that the environmental movement not always intervene negatively. I kept pushing and pushing trying to come up with some positive alternative whenever we'd intervene in a case.*

Berlin's partner in these strategy explorations was most often Charles J. Cicchetti, then an associate professor of economics at the University of Wisconsin, later chairman of the Wisconsin Public Service Commission. Berlin:

> *Finally we decided to shift our attention away from the supply side and focus on the demand side to see if that would help matters. I had endless discussions with Charley Cicchetti long into the night while he taught me about the theory of demand. . . . The idea [to go after rate structures] was probably pretty symbiotic with me pushing and pushing for some positive alternative and Charley probably at some point saying, "well if we don't focus on the supply side, the alternative is to go after demand."*
>
> *But that didn't solve it. We still had to work out an approach after that. There was a lot of beer consumed over the problem, I can remember that. . . . Ultimately, the Madison case was the result. . . .*

> *I remember presenting the idea of a rate design strategy, outlining a several years' approach, to the EDF's litigation committee which included Stu Udall and Lee White sometime in 1971. . . . The litigation committee approved the idea and we set out after Wisconsin.*

According to Berlin, Wisconsin was chosen as the trial state partly out of convenience and partly because of its reputation for progressive regulation which dates back to La Follette:

> *We looked around and it just seemed to be the best place to make our case at the time . . . primarily because that was where Charley was— and also Wisconsin had a chairman who was a young attorney who had a reputation as a progressive. Then, too, Wisconsin was small enough—the utility was small enough—and we thought that was important.*
>
> *Actually, it didn't remain small for long. The utility—Madison Gas and Electric—was soon joined by a number of other utilities as intervenors, and that's when the utilities brought in NERA [National Economic Research Associates].[27] That turned out to be good though because it focused a lot of attention on the case. . . .*
>
> *NERA . . . was brought in to do a rip job on Charley's story. They gave us some problems sure, but they were . . . good . . . economists. They basically helped clarify the value of marginal cost principles.*

Despite the EDF victory in Wisconsin, Berlin felt "in retrospect, it was probably a mistake to choose the Madison case to intervene in." The EDF had focused on time-of-day rates for residential consumers, not anticipating that the cost of metering would make a mandatory residential rate prohibitively expensive. Then, too, Madison Gas and Electric had a high load factor, which meant that savings due to load shifting would be minimal. Finally, the utility had few large industrial customers who could be metered; its customers were "nearly all residential and commercial."

The EDF's next step was New York and the Niagara-Mohawk case. Berlin recalled:

> *We recognized that if we were going to get any kind of national action on rate design we were going to have to do something in New York State. Niagara-Mohawk was the first case up and it had a lot of industrial customers. Actually we made another mistake in going into [that] case. . . . The utility had a high load factor—something like 70 per cent—and a winter peak [the state as a whole has a summer peak]. They still maintain that they will not be much helped by marginal cost pricing. This was the case that we brought in Bill Vickrey, who is like the father of marginal cost pricing.*

By the time it petitioned for the generic proceeding in New York, the Environmental Defense Fund had had considerable experience in promoting regulatory reform. The organization was quickly developing a national reputation for sponsoring responsible testimony. Its initial efforts had caught the attention of other consumer and environmental groups that were attracted to the possibilities of rate structure reform. At the time of the New York proceeding, the EDF was also intervening in rate cases in other states, notably California, and was soon joined in its advocacy of marginal cost pricing by witnesses from the Federal Energy Administration.

The Utilities

Like the EDF, New York's seven major investor-owned utilities petitioned for a generic hearing, but it did so for quite different reasons. Actually, the utilities were divided among themselves over the merits of marginal cost pricing. Some, such as Long Island Lighting Company (LILCO) and Con Ed, faced with huge capital needs and deteriorating load factors, were interested in the concept. Others, like Niagara-Mohawk, which still wanted to expand, were decidedly against abandoning the declining block rate.

That the utilities should enter a similar petition so soon after the EDF was a surprise to many of the New York commission's staff members, some of whom speculated freely about the utilities' motivations. Suggested one, "It could be that Irwin Stelzer [the president of NERA] got them all together to tell them about Kahn's general positions and to suggest that they do something because the commission would invariably move in this direction." Offered another, "I think the chairman gave them the message that rate design was on its way and invited them to petition for a generic hearing so that they wouldn't need to be dragged into the twenty-first century."

But the best insight into the utilities' motivations came from a source who was close to Alfred Kahn. When asked whether the utilities had been invited to petition for the hearing, the informant said:

In so many words, yes. Shortly after Chairman Kahn came on the scene he had a meeting with the chief executive officers of all of the utilities in his office in which he explained his commitment to marginal principles in rate design and his belief in the immediacy of the problem. He also sought to convince the utilities that the effort was worthy of their support and that the chairman recognized his need to rely on the assistance of the utilities—however reluctantly they would

*consent—for success. After due consideration the utilities agreed that
it was in their best interest to seek a generic hearing, so all of them
petitioned together.*

For some of the utilities, the generic proceeding represented a
means of slowing down the progress of rate structure reform rather
than speeding it up. Together, the utilities engaged NERA to present the affirmative case for marginal-cost-based rates, but individual utilities hoped the hearing would drag on without the commission taking action. In any event, the generic hearing safeguarded
utilities opposed to marginal cost pricing from having to deal with
the issue in the context of their own rate cases or from having to
follow the precedent of a marginal-cost-based rate established in the
rate case of a utility more favorably inclined to the concept. One of
Chairman Kahn's assistants explained the utilities' strategy as follows:

*First of all in a generic hearing there is no time deadline in which the
case must be heard and a decision reached. In our general rate cases,
by law we must issue a decision within eleven months of filing so that
the companies knew that a decision was going to have to be reached [if
it were heard in that context]. Without a deadline in the generic case,
it might be possible to drag the hearings out. . . . Some of the utilities
may have hoped that the commission might abandon its course in
marginal cost pricing, or that the commission membership might
change before changes could be instituted. Then, too, once a decision
has been reached to implement time-of-day rates there develops a
certain degree of acceptability [of] the notion by affected utilities and
consumers.*

The generic case was initially designed to take place in three
phases. In the first phase, the "threshold issue" of whether marginal
cost provided a reasonable basis upon which to design rates was to
be heard. Later phases were to consider lifeline rates and whether
individual utilities or the state as a whole should be the "proper
costing entity" for peak load rates—that is, whether peak periods
should be determined on a company-by-company or a state-wide
basis. On this last issue the interests of the utilities were clearly
fragmented. Niagara-Mohawk and other upstate utilities had a
winter peak, whereas the larger downstate utilities had a summer
peak. To Niagara-Mohawk, to be required to price its summer
power at peak rates on the theory that it was the growth in the state's
peak the regulatory commission should most be concerned about
rather than the peak of any single system was anathema. It would

probably worsen the utility's load factor and slow down its overall rate of growth. Whereas the utility supported the concept of a generic hearing on the threshold issue, it was vigorously opposed to anything other than a utility-by-utility determination on the "costing" issue. Was this an inconsistency? One of Kahn's assistants responded, "It's illogical on the surface, but not if your main objective is to slow down and maybe prevent the adoption of marginal cost principles. Then it's part of a clear strategy to delay."

When the threshold issue had been decided in favor of marginal cost pricing, the commission counted the yeas and nays that had been entered in the proceeding. In strong support of the concept were the Long Island Lighting Company, the Environmental Defense Fund, the Federal Energy Administration, the city of New York, and Chemung County Neighborhood Legal Services. In support, but less forcefully, were Consolidated Edison, the Central Hudson Co., the New York State Electric and Gas Co., and the Orange and Rockland Co.; the staff of the New York Public Service Commission; and the New York State Consumer Protection Board. In strenuous opposition were Niagara-Mohawk, Airco (a large industrial user), the General Services Administration, the Industrial Power Consumers Conference, the Multiple Intervenors (an industrial customer group), and the Rockland County Industrial Energy Users Association.[28]

Implementing Marginal Cost Pricing in LILCO's Rates: Developing External Support

The first utility to petition the New York commission for permission to restructure its electric rates in accordance with the principles of marginal cost was the Long Island Lighting Company (LILCO). The utility had an exceptionally poor load factor—45 per cent—and its summer peak had been growing rapidly. Its capital needs were huge, and it was unable to meet the interest coverage necessary to preserve the rating of its bonds. LILCO's managers hoped incremental pricing would help them deal with their problems, so they asked the commission to consider LILCO's petition in the context of the utility's 1975 general rate case. In deference to the other utilities, the commission agreed to postpone such consideration until after the threshold decision on marginal cost had been reached in the generic hearing. Within five months after the decision, LILCO had its new rate structure. In the interim, the commission, responding to external pressures, fashioned a policy for a number of the issues raised in the generic proceeding. It is neither possible nor necessary to render a complete account of the separate effects of each of these actors in designing and implementing a marginal-cost-based rate structure for LILCO. However, the roles of selected participants in the policy process are important enough to merit individual consideration.

The Governor's Office

Alfred Kahn was appointed to succeed Joseph Swidler by Malcolm Wilson, Rockefeller's lieutenant governor, who was defeated the following November by Hugh Carey. After Wilson appointed Kahn, he did not attempt to modify Kahn's preexisting views on marginal cost or on marginal cost's application to electric rate structures. When Carey became governor, he, too, made no real attempt to influence Kahn's thinking or his direction on rate structures. This absence of intervention prompted one member of the New York commission staff to observe, "The governor treated the commission with 'benign neglect' [on this issue]—which was good."

Rather than practicing benign neglect, however, Governor Carey was simply more concerned about other issues that touched New York's public service commission. These included the problems of siting new power plants, "intergovernmental relations" (a catch-all phrase used to describe a running battle with PASNY, the Power Authority of the State of New York), and constructing nuclear facilities and disposing of radioactive wastes. Time-of-day rates came fifth or sixth on the governor's list of priorities, according to staff members in the governor's office. In fact, when the governor interviewed a replacement for Kahn, one of his chief aides said, "I don't think they even talked about time-of-day issues."

Kahn understood the governor's priorities and earned the trust of his office through helpful consultation on the matters that most concerned him and his staff. Said one of the governor's assistants, "Fred worked closely with us on many of these things. . . . We'd send him over stuff . . . and he'd review it for us and be very helpful. He had a clear sense of [our] priorities, although time-of-day was his baby, of course." According to Carey's assistant, the problems of PASNY were among the problems on which Kahn was particularly helpful to the governor:

> *PASNY . . . was originally set up to build generating plants and transmission lines [to sell power to municipalities] which they did just great. But . . . we really don't know what to do with them. . . . They're just uncontrollable. We can't get them to do anything for us. You take a look at all the state environmental laws—there will be a caveat, "the above doesn't apply to PASNY." Water and air quality standards— they don't have to meet the same tests. They don't have to show need [for new construction] like other electric companies. . . . There are staff members who don't even return calls from the governor's office.*

Kahn was willing to provide both technical and political assistance to the governor's staff in dealing with the PASNY. Carey's assistant remarked, "Fred was very active and helpful to me on working with the governor to try to do something about PASNY. He was just as frustrated about the interaction with them as we were." As a result the governor and his staff did not attempt to interfere with Kahn's efforts on time-of-day rates. Carey's assistant observed, "We were so confident in what the commission was putting out under Fred, we just didn't worry about it. . . . We were pleased with what he was doing . . . [and] did not want to see him go [to Washington]."

The Legislature

There was never any serious legislative attempt to thwart time-of-day rates while Kahn was chairman of the New York State Public Service Commission, but some bills were introduced which would have strengthened the commission's movement in that direction. Rate structure reform appealed to the many legislators who were anxious to demonstrate concern about rising electricity prices. For similar reasons a number of bills were introduced to mandate lifeline rates of various sorts. None passed—in part because Kahn refused to push for the lifeline concept; in part because industrial and commercial users that could not stop time-of-day rates were adamant in their opposition to lifeline; and in part because consumer representatives failed to construct an effective coalition in support of lifeline.

The commission took no formal position on lifeline rates until August 30, 1978 (over a year after Kahn had departed for the CAB), at which time it concluded, in essence, that it would not order the general adoption of such rates without a specific mandate from the legislature. Kahn had anticipated that legislative order in September 1975 in a carefully worded statement given before the assembly's Committee on Corporations, Authorities and Commissions in which he had said:[29]

> *As long as we are told [by the legislature] that our obligation is to assure just and reasonable rates, to assure good quality service, to avoid undue discrimination, we know reasonably well what our obligations are. The terms are broad and, especially to a layman, vague; but they are terms of art, with a long history of interpretation. They mean, above all else, that total company revenue requirements are to be held as close as possible to cost, including a reasonable return on investment; and that the rates to different categories of customers should, similarly, be closely related to the respective costs of serving them. . . .*
>
> *The lifeline rate proposal is . . . a radical departure from those instructions, and I confess to a great deal of uncertainty about its desirability. . . .*
>
> *This is unfamiliar territory for a regulatory commission. I cannot tell how our commission will respond to the lifeline proposals that I am sure will come before us in the generic hearing on electric rate structure. . . . If we find, however, that these proposals are no more than a scheme to require that rates for a fixed amount of service to the poor and needy be subsidized by other consumers, we will have serious worries about whether we have the statutory authority or legislative mandate to engage in so extensive a scheme of social ratemaking, of income redistribution through our rate authority.*

Kahn's position on lifeline and especially his use of the buzz word "social ratemaking" appealed to three important constituencies— the utilities, the large users, and the commission's staff—and helped soften their opposition to time-of-day rates and in some ways won their support for such rates, which was what he really wanted to do.

Unlike the large users and to some extent the utilities, who opposed lifeline rates on purely materialistic grounds, the NYSPSC engineers objected to them because they offended their sense of mission. At the time the commission was considering lifeline proposals, the chief of the staff's rates and valuation section said:

> *Lifeline is still being considered and the commission hasn't had a recommended decision—but the staff's position in anticipation of those hearings [is] that lifeline [is] an explicit effort to build in subsidies into electricity rates which we don't think is justified; that is to say, [is] not our role to pass on.*
>
> *This is social policy and properly the role of the legislature. The mandate that the PSC has is to get out of the business of subsidies that characterized rates in the teens and twenties and begin to base rates*

more closely in line with costs. To the extent that lifeline deviates from that, it was not part of our mission.

Even the attempt to justify lifeline rates on the basis of inverse elasticity was dismissed. The section chief commented:

[Proponents] say that rates like that can still be economically efficient. I just don't know. What I'm more concerned with is that the residential consumer that you're dealing with is not a rational being. If he performed adequately all of the marginal decisions that you want him to, then maybe such rates could be justified. But I don't think he does. I don't know, for example, what the end block of Niagara-Mohawk is—despite the fact that that's the utility that provides my own power. What I know—what I react to and what other residential consumers react to, I think—is the total utility bill that I pay once a month. If you are charging people a kilowatt-hour rate of 30 cents and then giving them a $100 rebate each month, you may be able to preserve the proper signal in relative prices, but if people look instead at their total net bill, I'm not sure what the effect will be. . . .

What you want to do is set rates so that the customer who will make the proper marginal decision will do so. I think we can count on a large industrial user of electricity to do so—to compare the cost of production under various rate schemes and make the proper marginal choice. I'm less confident that a residential consumer who looks at the total bill will make the proper marginal choice even if rates embodying those signals are presented him.

The commission's director of research, an economist, agreed and expressed his concern that rebate proposals in keeping with inverse elasticity might threaten even modest hopes for basing rates on marginal cost, "One of the problems is [lifeline adherents] seize upon [incremental rates] as a possibility for redistributing income. I'm opposed to that. . . . [These] proposals just scare industrial and commercial users into opposition because they are afraid they'll end up footing the bill—and thereby jeopardize the whole move toward marginal cost."

Because of opposition from the chairman and NYSPSC staff and from the utilities and their large users, legislative proposals to mandate lifeline rates and other rate structure reforms were unsuccessful from 1974 to 1977. Staff assistants close to the chairman's office tended to discount legislative pressures on behalf of lifeline and even went so far as to express doubt about how seriously some sponsors supported their own measures. One staff assistant remarked:

Actually the most accurate description of the political environment is that there wasn't any. Oh, occasionally, we'd get some bill introduced in the legislature, but there was none of the arm bending "I appointed you and I can unappoint you" type of pressure that you might think. Mostly these were one-house bills. Someone would introduce a bill to limit the commission's discretion in rate setting, for example, or to mandate flat rates [rates that charged the same per kilowatt-hour charge, no matter how much electricity was used], or to outlaw declining block rates on the assumption, and even hoping, I think, that the other house would vote it down. We have a split party control of the legislature here, with each party controlling one of the sides. These bills quickly came to be discerned as really just grandstand efforts by individual politicians to get some favorable publicity back home. And since the newspapers back home were often not as sophisticated as the Albany press, it made a lot of guys look good to be out attacking the PSC.

Another commission source who dealt directly with legislators offered this insight:

When the legislature passes a law and the governor signs . . . then I don't consider that political. I consider it to be the will of the people and I'm willing to comply with it.

On the other hand, much of what the legislature does is not designed to pass a law, but to send us a message, and to send the folks back home a message. It's that kind of political pressure that we get from the legislature. In fact, I've had some members of the legislature tell me "don't worry, I'm not going to pass this, I just want you to know I'm very concerned about this so be sure to take a close look at this before you decide on something that may cause difficulty. . . ."

The problem is much of this legislature has a surface plausibility—we all want to control utilities and the best way to control them is to keep down rates. I've come to the conclusion that many legislators are not so interested in finding precisely the right response on a given issue as they are with finding a response that they can explain back home and get support for—rather than doing the additional study necessary to understand the more complicated issues— which, of course, would be more difficult to defend back home. Lifeline is one of these issues which has surface plausibility but is not cost justified.

The New York State Consumer Protection Board

Not everyone in New York state government was as favorably disposed toward the NYSPSC and its chairman as were the governor's staff and legislative leaders. In particular, Rosemary Pooler, the chairwoman and executive director of the New York Consumer Protection Board (CPB), carried on a running battle with Kahn. This personal animosity would be uninteresting if it had not affected the regulation of electric utilities, but it did.

The consumer protection board is formally comprised of the directors of several state agencies, including the chairman of the NYSPSC, but it has rarely met as such. As one CPB employee said, "The board is essentially a fiction. It has met only once in seven years. Our only accountability is to the chairwoman." Initially, the board was chartered to "represent the consumer interest" in, among other things, utility rate cases, and in 1977 it had a staff of four attorneys and three experts and resources of $350,000 to sponsor testimony in such cases. Roughly one third of those resources was devoted to electricity and gas cases.

At first, the staff of the NYSPSC supported the idea of intervention by the CPB "as a complement to what the commission itself was doing. We thought it would be a way to divide up labor . . . and greatly aid our own efforts . . . by bringing in the really high-priced and good experts, testifying on the commission's behalf." But in practice, things did not work out that way. NYSPSC staff members became highly critical of CPB-sponsored testimony and they questioned the value of consumer protection provided by agencies other than their own. For their own part, CPB employees contended that the NYSPSC was simply jealous of "our position as another government agency, presumably usurping their bailiwick." Said one, "Typically, we take positions to the left of the PSC staff. The chairwoman perceives it as her job to keep the pressure on the commission. It's political, and whether or not her methods are appropriate, I don't know."

Asked about the NYSPSC, Rosemary Pooler responded simply that the NYSPSC had "a lot of older staff people . . . that are out of touch" and that Alfred Kahn was "a brilliant man, but . . . an intellectual aristocrat [who was] not very good at politics." Asked to explain, Pooler said of Kahn:

> *He knew the business of regulating . . . long before the rest of us. But he made some bad political judgments that cost him some points. I'll give you one example—a small thing, but it illustrates his general behavior. Before he was named chairman he had done some consulting for AT&T. He was continually bothered by what he thought were unjust conflict of interest charges. He didn't appreciate its political dimensions. . . .*
>
> *He was often bothered by that lack of political understanding. It affected our relationship and hurt him. He shouldn't have done that. We were natural allies—protecting the consumer.*

If the NYSPSC-CPB alliance broke down in general rate cases, it was even more likely to falter on questions of rate structure, for while the NYSPSC staff was strongly opposed to lifeline rates the various community action groups that formed the clientele of the CPB were lifeline advocates. But if confrontation threatened, it never developed. "We've learned that marginal cost is the way to go to get something done," said Pooler. "We've been working with the consumer groups on that. . . . We've received $200,000 from the Department of Energy [and] we're using that money to work with consumer groups—training sessions; help them sponsor testimony. They're coming along because they see that if we don't put that coalition together . . . we won't get anything done, and that substance is more important than ideological purity." Asked if she meant that the CPB would be willing to give up lifeline, Pooler said:

Absolutely! We sponsored the testimony of Cicchetti in the generic case. It was a tepid lifeline testimony. That came right after I came on as chairman, and I supported it then. I didn't know anything about rate structures then, but I've learned since. I've had to confirm my support for that statement, and I've done it. We said that it was a "lifeline" testimony then only because the word was convenient. We're really looking for rate reform. It's a matter of getting power to do something.

If anything, the consumer protection board wanted to move even faster than the public service commission thought prudent toward adopting time-of-day electric rates. A CPB employee recalled meeting with representatives of the PSC and the governor's office in the fall of 1976 to decide on whether to introduce a bill that would mandate action by companies on time-of-day rates as part of the governor's legislative package. Of the bill, the CPB employee said, "It was withdrawn as a compromise when Kahn said he was doing all he could."

Fragmenting the Opposition: Seeking Cooperation through Compromise

With support for time-of-day rates from the governor, the legislature, the consumer protection board, and at least some of the utilities, the only unified opposition to marginal-cost-based rates came from large electric users, who feared the proposed rates would cause them to pay more for their electricity. To block the adoption of marginal cost pricing, the large users acted through both the formal and the informal regulatory processes.

Large industrial and commercial users' formal opposition to marginal cost pricing was centered in the generic hearing, where they challenged the theoretical justification of marginal cost concept and attempted to cast doubt on evidence presented in support of it from Great Britain and France. Large users hoped second-best considerations would force the commission to abandon its effort, but such was not the case. The commission effectively closed discussion on the marginal cost issue by placing the burden of proof on its opponents. In defense of that action, Alfred Kahn wrote, "What [opponents] neglect to observe is that this consideration applies *equally* to leaving the price of electricity *where it is*: the principle of 'second best' raises questions about the validity of basing electricity rates on accounting costs, too."[30]

The industrial consumers' angry response to Kahn's action was to file motions with the commission asserting that Kahn had determined in advance of the generic hearing his own answers to the issues that were to be raised in the hearing. The attorney for one industrial group berated the final decision on marginal cost, "At times I felt Kahn should have just issued copies of his text [*The Economics of Regulation*] as the decision—perhaps with the hard cover replaced by a soft one. The hearings produced nothing. They were simply an expensive seminar designed to give us Fred Kahn's view."

The large users' informal opposition to marginal cost pricing was more effective in shaping the final policy of the commission than their formal opposition. In fact, it was in response to political pressure from large users that Kahn devised a policy that shattered the unity of the large users' opposition, answered their charge that a new rate solely for large users was discriminatory, and secured the success of marginal cost pricing, even though it muted its impact. Under Kahn's strategy, which was as simple as it was effective, no consumer class would be required to pay more under marginal-cost-based rates than it had under traditional ratesetting methods and in fact might pay less.

Effects of Kahn's Strategy

Essentially, Kahn's policy decision meant that no excess revenues would be rebated to residential customers even if such rebates were justified by virtue of inverse elasticity or because industrial users took a greater percentage of their electricity on-peak rather than off-peak. Furthermore, the policy meant that if the load factor of users who were metered for time-of-day rates improved, their contribution to the utility's overall revenue requirement might be reduced. Even more important, it meant that if some large users ended up paying more under the new rates, others would necessarily end up paying less. Finally, the policy solved a troubling bureaucratic problem for Kahn—namely, the problem of what to do about the fully allocated cost study Joseph Swidler had authorized and the commission staff had put much effort into. Ron Liberty, chief of the commission's rates and valuation section, explained how the policy was fashioned:

> *It was in one of those sessions with Fred. We spent hours talking about these issues—how to design rates that would do what he wanted to do. I'm sure that it was on one of these occasions when we were talking about revenue requirements and what to do with excess revenues that the idea was developed. It just struck us as satisfying a number of concerns—what to do about the fully embedded [allocated] cost study, how to set the revenue requirement, and most importantly, its effect in answering the discrimination issue and in [mollifying] the opposition of the large industrial users upon whom the LILCO experiment was going to be made.*

If Kahn's policy was politically appealing, it was not inspired by principles of economic efficiency, a fact not missed by the commission's chief economist, Robert Smiley. Smiley recalled an early strategy session with Liberty and others over the lifeline issue:

> *Basically, we all decided that the goal of redistribution ought to be decided in the legislature and that it would be wrong for us to do anything other than to try to get prices to conform to cost.*
>
> *That issue later came up in the LILCO case where we decided to keep the revenue requirements by class the same as under the fully embedded cost approach. I pointed out that that was just as arbitrary as a lifeline approach which we had just criticized—that it assumed authority to make the kind of judgments that in the lifeline case could only be made by the legislature. . . .*
>
> *[Their response was] . . . "Oh."*

But Liberty refused to accede to Smiley's point of view so easily. Liberty stated:

> *I guess to the extent that total class revenue requirements are less than they would be if all rates were based on marginal cost, [the policy] would constitute a subsidy. But you have to draw the line somewhere.*

There is legal and there is economic discrimination, and we felt that this economic discrimination was justified on other grounds. In the LILCO case we were only making one group meter and we felt in order to avoid the discrimination issue we would have to assure the large industrial users that as a class they would not have to pay any more under this scheme. Then, too, we had just completed the fully allocated cost study, and on that basis we knew what revenue requirement to set by class—but we had no idea on the marginal cost basis.

Most important, Kahn's proposal promised to turn what had been a solid block of opposition into a group of varied interests, some of which would benefit from the proposal:

[The policy] was definitely necessary as a political strategy for gaining acceptance of the concept. The industrial users love this decision. And it was important that we assure them that they could depend on this approach for the relevant future—in all of the rate cases to come . . . based on the generic proceeding. They don't want to let us forget this because their great fear is that marginal cost will be used to adopt rates that will make the large customers end up footing the bill for residential consumers.

The Meeting of July 8, 1976

Kahn and his staff soon had a chance to try out their new policy on a group of industrial users who met with Kahn on July 8, 1976, a month before the commission issued its decision in the generic hearing, to protest the move toward time-of-day rates. The meeting was arranged by John Dyson, commissioner of the state's commerce department, to whom the large electric consumers had turned for help when it became evident that they were not winning the battle against marginal cost pricing in the formal regulatory process. Dyson had a reputation as an outspoken advocate of economic recovery and industrial development in New York and he reportedly enjoyed a close association with Governor Carey. Dyson was known to have attacked the Department of Environmental Conservation and the Department of Education for imposing undue burdens on business and for inefficiency and had received a great deal of press attention and business support as a result. Dyson's standing was further increased when, shortly after these attacks, the head of the Department of Environmental Conservation was fired by the governor and the Chancellor of Education was removed by the Board of Regents. Industrial users were searching for any advantage that might deter

Kahn from his well-known intentions and hoped that Dyson would provide that advantage by attacking Kahn and the NYSPSC as he had earlier challenged the departments of environmental conservation and education.

In their hope the industrial users were disappointed. In February, Dyson's staff briefed him on the activities and direction of the NYSPSC under Chairman Kahn. Early in March, Kahn and Dyson met for lunch, at which time Dyson was thoroughly persuaded by Kahn's defense of economically efficient rates. The commerce commissioner became convinced that industrialists were suffering from misinformation and misunderstanding regarding Kahn's intentions at the NYSPSC. On April 1, 1976, in a letter to Kahn, Dyson suggested that he would be happy to arrange a meeting for Kahn with industrial customers. Kahn agreed to the meeting, and in early June invitations were sent out to all of the business leaders who had registered their distress with Dyson about the goings-on at the NYSPSC, including some who were pressing for Kahn's impeachment.[31]

Nearly forty people attended the meeting arranged by Dyson, which was held in Kahn's office in July. The session was a great success for Kahn. He discussed the principles guiding the commission's efforts and then shared with the business leaders the commission's proposed policy to keep class revenue requirements unchanged, explaining that some of them could gain by the time-of-day proposal and that the average user would be unaffected by it. As one would expect, the responses of the business leaders varied dramatically. From one of Kahn's assistants came this observation:

Fred's handling of that meeting was simply a delight to watch. It convinced me that there were probably not more than six people in the country that could have done for New York what he did. Sure there are a lot of fine economists and micro theorists, but he was able so many times to hit the nail on the head in response to particular questions and thus gain support that he otherwise would not have received.

I wish I could give an example that would do him justice. In the meeting at one point, I forget the exact context in which this came up, somebody asked whether or not there could be other values than efficiency taken into account in setting prices, like fairness. Fred said something to the effect that we've got to set prices some way, and that setting them in accordance with cost seems to be the most logical way to do it. If you want to argue for something else you're going to lose.

He said that—he, simply with fist in hand, said that—"you're going to lose." Then he posited the question, "Why?" and answered, "be-

cause if prices are set in some fashion other than on the basis of cost, then it's the political process which will set prices and the political process asks who will gain and who will lose. And let's see who's going to gain? Schools will gain, hospitals will gain, the small business user will gain, and the residential customer will gain. And who will lose? You are going to lose."

Then he threw in some empirical political behavior models stuff that said that households had the votes . . . and said it's not the little guy but the industrialists that they are going to sock it to.

"Why," he said, "you ought to stand up and cheer this proposal, because it is the only thing that will save you."

Not everyone cheered. Users that could shift their consumption to off-peak hours would benefit under the proposal, but those that could not would end up paying more and they were not happy about that prospect. This was especially troubling to large retailers that operated large shopping malls and department stores, who felt that the nature of their businesses prohibited off-peak consumption. But their argument that such rates would drive them out of business did not impress Kahn's staff or John Dyson. One of Kahn's colleagues remarked, "The retail merchants association has continued to be actively opposed to time-of-day rates. Actually, the thought occurred to me [during the meeting]—although I immediately dismissed it as inappropriate—that these are exactly the customers which should be priced at full marginal cost, because unlike industrial users, their threat to run off to Utah or someplace with a more favorable regulatory climate is less credible."

Gary Perkinson, the executive director of the New York Council of Retail Merchants, was one of those in attendance at the meeting who was outraged by Kahn's presentation, "Kahn wrote the book on regulation and then spent the rest of his life trying to prove it. He started the meeting by saying, in short, that he couldn't understand why businessmen are so stupid they can't see that this is in their best interests. It went downhill from there."

During the meeting, Perkinson engaged Dyson in an argument, demanding to know who was representing the consumer interests of the retail merchants who were members of his association. Finally, according to one of those in attendance, Dyson said: "Gary, if you want to sell nylons in Poughkeepsie, you can't do it from Connecticut. I have to be more concerned about industries that might leave and are leaving the state. Retail users are captive." To which Perkinson reportedly replied, "I can't believe I'm listening to the Commissioner of Commerce telling me he doesn't care about the retail users in New York."

On balance, however, the meeting ensured the success of Kahn's policy initiative. A source close to Commissioner Dyson summed up the meeting's effect in the following way:

> *I have the greatest respect for Kahn as an economist who can make obtuse principles understandable. The meeting strengthened my opinion of Kahn because of industry's reaction. A few of the industries appeared to be honestly surprised. Kahn's presentation hit them like a shock: "Why am I fighting this way when I may make out on it?" The industries there came away impressed. I almost believe some were turned around on the basis of hearing Kahn.*
>
> *More than anything else these people learned that John Dyson wasn't going to attack the PSC, too, because he was convinced by Kahn.*

The Formulation and Effects of the LILCO Rates

Having fashioned a policy that shattered opposition to a new marginal-cost-based rate structure and having won the support of key people in the executive branch and in the legislature, the New York State commission was able to dispose of remaining issues and dispel further opposition by adopting a policy of gradualism. The tariffs approved in the LILCO case (see Table 4.1) applied to only the utility's largest customers, those whose monthly recorded demand exceeded 750 kilowatts in any two months of the previous year. The daily peak period was broadly defined as 10:00 A.M. to 10:00 P.M. to avoid the problem of shifting peaks; similarly, the seasonal peak was broadly defined as June through September. The staff recommended that the needle peak problem be solved through interruptible rates and "load shedding" devices rather than through further refinement of peak load rates.

In formulating the LILCO rates, it was in defining marginal costs that perhaps the greatest compromises were made. Using a methodology developed by Ralph Turvey, a British economist, LILCO rate engineers identified three periods: peak, intermediate (or shoulder peak), and off-peak. No capacity charge was assigned to the off-peak period on the assumption that there was a zero probability that that period would eventually become the peak. Upon the first application of the methodology, the ratio between the peak and intermedi-

Table 4.1 LILCO's Time-of-Day Rates for Customers With Demand in Excess of 750 kw in Any Two Months

	RATE PERIODS		
	1	*2*	*3*
	Off-Peak	*On-Peak*	*Intermediate*
Per Meter, Per Month	12:00 Midnight to 7:00 A.M.	June to September Inclusive, Except Sundays 10:00 A.M. to 10:00 P.M.	All Remaining Hours
Demand Charge per kwh:			
Secondary	None	$6.40	$1.60
Primary	None	5.60	1.40
Transmission	None	4.20	1.05
Energy Charge per kwh:			
Secondary	1.73	3.25	2.67
Primary	1.71	3.18	2.61
	1.66	3.02	2.52

Source: Long Island Lighting Company Service Classification No. 2-MRP, Large General and Industrial Service With Multiple-Rate Periods (December 29, 1976).

ate periods was in the range of 18:1 to 20:1.[32] LILCO felt that such a ratio represented too abrupt a change from its traditional rate structure and would cause hardship to peak customers, so it recommended that the commission approve instead an 8:1 differential. The commission was in such complete agreement with the spirit of this recommendation that it actually authorized a demand charge of only $6.40 per kilowatt for peak use but $1.60 per kilowatt for shoulder peak use. This was a differential of only 4:1, certainly not what anyone could call a strict marginal cost assessment.

At the same time LILCO's energy charges were determined on the basis of short-run marginal cost on the assumption that oil would remain the utility's marginal fuel source in the foreseeable future and because it was felt that for strategic political and economic reasons New York State should not have a policy that encouraged the use of oil by underpricing utility operating costs. The customer charge for billing and other services was eliminated entirely.

The compromises made in defining marginal costs and in keeping customer revenue responsibilities unchanged by class had the effect of creating a new rate structure that did not differ greatly from a peak-load pricing scheme based on average historical costs, a fact that distressed some historical cost proponents, who would like people to stop calling the LILCO decision a milestone for marginal cost pricing. But staff engineers of the New York commission thought the compromises were necessary to take the first step toward marginal cost pricing and were quite willing to admit that LILCO's was not the ideal rate structure. One staff engineer commented:

> *As it turns out you could have taken the marginal cost capacity study, not even used it, and come up with the same charges based on embedded [historical] costs. . . .*
>
> *We knew we were conservative. Without the study we could have guessed, but we wouldn't have known if we were conservative or not. We wouldn't have had a way to defend it. "Marginalists" were happy [with the decision] and average cost supporters were happy.*

A year after the new rates went into effect, LILCO's preliminary studies showed that the rate structure was having exactly the effect Kahn had told industrial users it would have. Of the 173 accounts for which the NYSPSC staff had figures, only 11 had paid more under the new rate structure than they would have paid under the traditional one. All of the increases were less than 2 per cent whereas, among those who had received decreases, savings had ranged up to 21.3 per cent. As a class, the 173 large users had saved over $1.5 million, about 4 per cent.

Conclusion

The effort of the New York State Public Service Commission under Chairman Alfred Kahn to adopt and implement a new rate structure for electricity based on the principles of marginal cost represents a landmark in the history of public utility regulation. Other state commissions have looked and will look to New York for leadership in reforming their own rate structures. Many will face the same issues that were raised in New York: whom to meter; whether to use marginal cost pricing and how to define marginal costs; what to do about excess revenues; and, especially, whether to adopt lifeline rates. Not all of these issues will be decided in the same way in each state. Bureaucratic and environmental constraints differ, as do the objectives of regulatory commissioners and utility executives. California, a state whose commission is more like New York's than any other's, adopted a lifeline rate despite what we have seen to be solid opposition to the concept in New York by its commission's staff.

This difference in policy prompts us to ask why California chose a lifeline rate, a question to which we turn in Chapter 5.

But there are other reasons that make the New York case a compelling study, and they have to do with examining the effects of structural changes in politics and economics on regulatory outcomes. Would Alfred Kahn have been able to pursue a "politics of efficiency" with as much success in 1964 as he did in 1974? The answer is almost certainly no. A decade earlier Kahn's enthusiasm and skilled advocacy of the marginal cost concept would have at best been treated with indifference. The real price of electricity was falling, and utilities were anxious to promote the growth of consumption. Any policy that threatened those plans would have faced unrelenting opposition. Nor would there have been any support for such a proposal except from a handful of academics whose occasional journal articles on the subject lamented a world in which marginal cost received too little attention. Then, too, as long as marginal cost was shifting downward there would not have been excess returns but rather a shortfall of earnings if rates had been based on marginal cost. Utilities would have been unable to recover their fixed costs and would have been allowed to charge some consumers more than marginal cost to make up the difference. In that case the inverse elasticity rule would have worked to the detriment of residential users. Since their demands would have been the most inelastic, their rates would have received the greatest increase above marginal cost. This would certainly have provoked a good deal of consumer protest. In political terms in 1964 the declining block rate structure made sense; in economic terms it was not terribly far off base because it promoted growth and economies of scale.

In the aftermath of technological change, inflation, the environmental movement, and OPEC, however, all of this was changed. Some utilities no longer viewed growth as their chief objective because scale economies had been largely exhausted. Environmentalists certainly did not want growth. No one—least of all residential consumers—wanted rate increases. The opportunity (perhaps necessity) for change was created. Into this changed environment stepped Alfred Kahn and other economists, armed with a set of logically consistent principles that promised to solve the problems of a diverse constituency: the NYSPSC staff, which was worried about revenue erosion; LILCO, which was worried about its capital requirements; the Environmental Defense Fund and the Federal Energy Administration, which were worried about capacity growth and fuel conservation; and residential consumers, who felt that traditional pricing methods discriminated in favor of the electric utilities' large users.

Alfred Kahn left the New York Public Service Commission in June 1977 to serve as chairman of the Civil Aeronautics Board. By then the reform of electric rate structures was on the President's agenda as well as on that of most state public utility commissions. Assessing Kahn's individual contribution to the reform movement is a difficult thing to do. It is easy to overstate his influence because his style and intellect thoroughly dominated the commission during the three years he served as chairman. Yet it seems clear that without the structural changes that occurred in the political economy of electric utility regulation during the ten years that preceded his chairmanship Kahn could have done little, just as it seems clear that those same structural changes will make it difficult to distinguish Kahn's personal effect on the reform movement a generation from now. Perhaps the best assessment of Kahn came from one of his staff assistants:

> *Not every commission can break ground as New York did under Fred. The important [condition] for breaking ground is that you have to be good. . . .*
>
> *It was one of those happy chances in history when the man and the opportunity met. . . .*
>
> *[But] in thirty years I don't think it will make much difference whether Kahn . . . had ever lived because we will be where we would have been otherwise, anyway. Marginal cost now has a number of proponents in other states and the staff at the FEA is solidly convinced by it and really pushing hard for it. [It was included] in the President's energy message, so events are pretty well moving us in that direction.*

Endnotes

1. State of New York Public Service Commission, Order Instituting Proceeding, Case No. 26806 (January 29, 1975). A generic rate proceeding is one which examines ratemaking principles outside the confines of any utility's specific rate case. Hereafter, the State of New York Public Service Commission will be referred to as NYSPSC.
2. Ibid.
3. NYSPSC, Opinion and Order Determining Relevance of Marginal Costs to Electric Rate Structures, Case No. 26806, Opinion No. 76-15 (August 10, 1976), pp. 31, 34.
4. NYSPSC, Opinion and Order Requiring the Establishment of Time-of-Day

Rates for Large Commercial and Industrial Customers, Case No. 26887, Order No. 76-26 (December 16, 1976). On June 27, 1977 the New York State Council of Retail Merchants and several other parties filed a joint petition with the supreme court of the state of New York, county of Albany, to annul the above decision on the grounds that the selective application of a marginal-cost-based rate amounted to unjust discrimination against the utility's largest customers. In May 1978 the court held in favor of the petitioners, and LILCO and the NYSPSC are in the process of appealing its decision before the New York State Court of Appeals. The case is being closely watched by students of rate reform throughout the country.

5. Alfred E. Kahn, *The Economics of Regulation: Principles and Institutions*, 2 vols. (New York: John Wiley & Sons, 1970–1971).

6. *Electrical Week* (February 17, 1975), p. 7.

7. *See Business Week*, May 25, 1974, p. 102.

8. The rate experts in the communications division were not difficult at all to convince, having been introduced to time-of-day rates by the Bell system. Many of Kahn's greatest contributions in New York related to his efforts at telephone regulation, but they are not the focus of this study.

9. Motion to Recuse on Behalf of Multiple Intervenors, NYSPSC Case No. 26806. To recuse a judge is to challenge the judge as prejudiced or otherwise incompetent to act. The motion was filed by the law firm of DeGraff, Foy, Conway, & Holt-Harris.

10. NYSPSC, Decision of Chairman Kahn on Motions to Recuse, Case No. 26806.

11. NYSPSC Interoffice Memorandum (July 26, 1972).

12. Kahn, *Economics of Regulation*, vol. 1, p. 182.

13. Alfred E. Kahn, "Utility Rate Regulation: Applications of Economics" (Paper presented before the American Economic Association, Atlantic City, New Jersey, September 17, 1976) p. 23.

14. Ibid., pp. 23–24.

15. *Wall Street Journal*, September 21, 1978, p. 1. NYSPSC staff engineers estimate that the additional reading and billing costs of a two-dial meter are negligible (DDA interview, November 1978).

16. For Kahn's view see *Economics of Regulation*, vol. 1, pp. 195–98.

17. Public Service Commission of Wisconsin, Application of Madison Gas and Electric Co. for Authority to Increase Its Electric and Gas Rates, 2-U-7423 (August 8, 1974), p. 49. The analytical solution to the shifting peak problem is simple. Both periods represent the peak, so both should share the costs on the basis of the respective intensities and elasticities of demand. It is the practical problem of designing a rate that varies with the ever-shifting balances of capacity and demand that is at issue here. *See* Kahn, *Economics of Regulation*, vol. 1, pp. 91–94.

18. Kahn, *Economics of Regulation*, vol. 1, pp. 70–76, 83–86.

19. This is not to suggest that an alternative allocation method, one perhaps based on average cost, would involve any less judgment. Any nonmarket allocation process is essentially judgmental. But the example does illustrate the kind of issues that emerge in translating marginal cost principles into rates.

20. Kahn, *Economics of Regulation*, vol. 1, pp. 89–109.

21. Arthur D. Little, Inc., *Topic 9: Mechanical Controls and Penalty Pricing* (Cam-

bridge, Mass.: Electric Utility Rate Design Study, January 15, 1977), pp. 9–22, 23.

22. The "Ramsey rule" takes its name from Frank Ramsey, who first established its efficiency properties. *See* Frank Ramsey, "A Contribution to the Theory of Taxation," *Economic Journal* (March 1927), pp. 4–61.

23. *See* J. W. Wilson & Associates, *Elasticity of Demand: Topic 2* (Washington, D.C.: Electric Utility Rate Design Study, 1977).

24. *See* Robert H. Frank, "Lifeline Proposals and Economic Efficiency Requirements," *Public Utilities Fortnightly*, May 26, 1977, pp. 11–15.

25. Public Service Commission of Wisconsin, Application of Madison Gas and Electric Co.

26. Edward Berlin, Charles J. Cicchetti, William J. Gillen, *Perspective on Power* (Cambridge, Mass.: Ballinger, 1974).

27. NERA is New York-based and headed by Irwin Stelzer, who received his Ph.D. under Kahn at Cornell.

28. NYSPSC, Opinion and Order Determining Relevance of Marginal Costs to Electric Rate Structures, pp. 4–6.

29. Alfred Kahn, "Statement of Alfred Kahn on Lifeline Electric Rates Before the Assembly Committee on Corporations, Authorities, and Commissions" (September 24, 1975), pp. 2–4.

30. *See* his statement in Robert G. Uhler, *Rate Design and Load Control* (Palo Alto, Calif.: Electric Utility Rate Design Study, 1977), p. 152.

31. Letters on file at the New York Public Service Commission.

32. NYSPSC, Opinion and Order Requiring the Establishment of Time-of-Day Rates, pp. 4–6.

Chapter 5

THE POLITICS OF PRICING: LIFELINE RATES AND REGULATORY LAG IN CALIFORNIA

To suggest that the California Public Utilities Commission (CPUC) does not enjoy the esteem of the industrial and utility interests it regulates would be to put it mildly. The state commission is one of a half dozen often mentioned by financial analysts as fostering an "un-favorable regulatory climate." In 1978 *The Wall Street Journal* editorialized that the commission's majority "has made it a positive villain to utility executives and shareholders."[1] But the agency has not always been viewed as such. Indeed, under Governor Ronald Reagan the CPUC was characterized in the press as the "give-away" commission, run by inept political hacks and staffed by demoralized bureaucrats. So favorable toward industry and the utilities was the commission that one California utility executive conceded that it was "like a friendly bear in a china shop" and was causing the company public-relations problems.

This is the story of how a state utility commission changed from industry's servant to industry's villain. The heart of the story lies in the September 1975 decision by the California PUC to order Pacific Gas and Electric Company, California's largest supplier of electricity, to adopt lifeline rates for its residential customers despite widespread opposition to these rates by the state's utilities, their largest customers and the CPUC staff. One week after the PG & E order,

on September 23, 1975, Governor Jerry Brown signed into law a bill
that mandated lifeline rates. Most informed observers agree that
this bill would not have passed the legislature had it not been for
the efforts of two Brown-appointed CPUC commissioners, Leonard
Ross and Robert Batinovich.[2]

In this chapter we will take a close look at the CPUC's lifeline
decision and at the political environment in which it was reached.
Like the New York State Public Service Commission, the California
commission is a leader among state commissions restructuring electric rates. It, too, has adopted marginal-cost-based, peak load rates
for large industrial and commercial users of electricity. Its lifeline
law, however, is unique among states. By 1980 no other state legislature had ordered a utility commission to adopt such rates, but at
least a dozen commissions were experimenting with lifeline rates on
their own initiative. Most states are now experiencing some of the
same pressures and are having to deal with some of the same issues
that surfaced in California—an important reason for investigating
the California lifeline case.

There is another—and for our purposes more important—reason
for investigating the California lifeline case, however. In contrast to
the New York case, the California case provides us with an opportunity to evaluate the determinants of regulatory behavior when the
regulatory task involves the making of a Type III loosely constrained
point decision. In Chapter 1, point decisions were characterized as
policy-setting tasks involving a single decision of the yes-no variety
which allocate scarce values or impose burdens on one of a number of rivals. As we shall see, lifeline rates confront commissions
with just such a decision. The task facing the commission is simply
one of casting a vote for or against the policy. Once the vote is cast,
the issue can effectively be ignored.

In order to comprehend why the California commission, unlike
the New York commission, embraced lifeline we will need to devote relatively more attention to the CPUC's external environment.
It will be important to show how and why this environment changed
over time, how new client groups formed, and how the commission
responded to changed external constraints.

An analysis of the CPUC's environment is an important undertaking if only because, as we reasoned in Chapter 1, the entrepreneurial behavior of a regulatory executive faced with an unconstrained point decision is best understood in terms of his attempts to

forge a coalition of support from among external client groups. In this case, however, an evaluation of the changing regulatory setting allows us to accomplish another purpose as well. One of the crucial developments in the series of events leading up to the lifeline decision was the ill-conceived attempt of an earlier commission president, John P. Vukasin, to facilitate rate increases for utilities by reducing regulatory lag. As we shall see, Vukasin acted as if his initiative alone were sufficient to reduce regulatory lag. He did not appreciate until it was too late that he needed to depend on the cooperation of the commission staff to achieve his goal. Vukasin's misguided efforts resulted in great notoriety for the CPUC and sparked the organization of anticommission consumer groups that later proposed lifeline. Vukasin's efforts are interesting because they illustrate the conflict-minimizing behavior of regulatory executives faced with tightly constrained planning tasks, a Type II case in the taxonomy of regulatory situations. For this reason the case study of the California lifeline decision begins with an evaluation of the CPUC under John Vukasin.

Reducing Regulatory Lag: The Type II Case of Conflict Minimization

The California Public Utilities Commission, as noted in Chapter 2, was established at the beginning of the 20th century by a reform governor, Hiram Johnson, to regulate the railroads and California's growing utilities. The commission's headquarters were located in San Francisco instead of Sacramento, ostensibly to isolate the commission from political pressure. The commission's five members are appointed to six-year staggered terms by the governor, and its decisions may be reviewed only by the state supreme court. From its origin, the California commission and its staff took pride in attracting "the best engineers Berkeley was training."[3] The commission has usually taken an activist stand on the leading issues of the day. During the 1960's, when electric utility costs were falling and commissions throughout the country seemed content to allow rates that had been fixed at an earlier, costlier time to generate record earnings for utilities (as long as the rates didn't become a political issue), the California commission staff, with the backing of the commission, initiated and achieved several large rate reductions from the state's major utilities. The staff's zeal for negotiating rate reductions was not limited to electric or gas utilities, however. In a noted telephone case, the commission ordered Pacific Telephone and Telegraph Company not only to reduce rates but to refund millions of dollars of overcollections. Although the state supreme court overturned the commission's rulings on refunds as "retroactive ratemaking," the court's decision did not stop the commission from pushing for low rates in other telephone cases, a course of action that signaled the commission's willingness to punish companies that refused to cooperate on rate reductions.[4]

While their costs were falling, the utilities acceded, albeit grudgingly, to the commission's activist stand. But in the late sixties, when they saw their costs rising rapidly, the utilities decided to take a more activist role in the political process themselves. They contributed heavily to the gubernatorial campaign of Ronald Reagan in 1966 in the hope that Reagan would appoint a commission more sympathetic to their concerns.

The utilities wanted two things: higher allowed earnings and a reduction in regulatory lag. Governor Reagan wasted no time in attempting to fulfill their demands, and as he did so external constraints exerted by utilities on commission decision-making began to tighten. Shortly after his election Reagan publicly announced his support for a rate increase for the state's largest utility, despite the fact that the commission had not yet heard the utility's case. Later, in 1969–1970, when the Reagan-appointed majority was finally in place on the commission, the commission's approval of utility rate requests began to show a much more favorable disposition toward the utilities—so much more so, in fact, that one disgruntled staff attorney said, "All of a sudden there were many more people from the industry walking around here and getting inside to see the commissioners. The Reagan people would take the utilities to task over little cases, but when it came to the big cases they were there with the wheelbarrow."

Spurred by the magnitude and frequency of rate increases, the press took the CPUC off the financial page and put it on the front page. Negative publicity about the CPUC reached a head in 1974, Governor Reagan's last year in office. The state's largest daily newspaper and at least one television station featured lengthy exposés of the CPUC's personnel and procedures. The *Los Angeles Times* conducted a ten-week investigation.[5] Television station KGO in San

Francisco serialized a highly critical report on the CPUC's internal workings. The report, which the station said was based on a year-long study, was broadcast on the evening news in eleven segments. The adverse publicity caused a chain reaction: The press reports that the commission was mismanaged and pro-utility sparked the formation of consumer groups. When these groups clambered noisily into commission hearing rooms, the press covered them in full. When their demands for commission documents or for admission to commission conferences were turned down, the groups charged—and the press reported—that the commission was secretive and unresponsive. One newsman went so far as to stage his own sit-in at a commission conference. Naturally, for the benefit of the public, he had arranged for a photographer to film his ejection.[6]

John P. Vukasin's Fight to Reduce Regulatory Lag

It was the enormity of the rate increases permitted by the CPUC that ignited the media, but it was a series of commission-initiated procedural and administrative reforms designed to reduce regulatory lag that added fuel to the flames. These reforms were facilitated by the passage in 1969 of a law that altered the structure of the CPUC to make the president of the commission the agency's chief administrative officer. Under the new law, the president was elected by his colleagues from among the five commissioners appointed by the governor. Prior to the law's adoption, the chief administrative officer of the commission had been its executive secretary; in 1969 the executive secretary was William Dunlop, the brother of former U.S. Labor Secretary John Dunlop. John P. Vukasin became the first CPUC president to enjoy the position's widened authority, which included the power to determine staff assignments.[7]

Before embarking on major procedural changes, Vukasin and his commission colleagues sought to strengthen their position with the commission's staff. One of their first successes in this regard came when Dunlop, the executive secretary, left to take a similar job at the state board of equalization. Dunlop had spent nearly his entire career at the CPUC, joining the agency in the 1930's. He had worked his way up through the ranks, serving as everything from beginning rate engineer to hearing examiner. With Dunlop gone, no single individual could command the loyalty of the staff in all divisions. One by one the remaining division heads were either won over to support for the new commissioners or enticed to leave the CPUC staff. The commission's chief counsel, Mary Moran Pajalich, who headed the 25-member legal division, sometimes proved an embarrassment to the commission by publicly opposing its positions, but she finally left when Governor Reagan appointed her to the San Francisco Municipal Court. Both Pajalich and Dunlop were replaced by people from outside the commission.

Administrative and Procedural Changes at the CPUC. Vukasin made no secret of his concern for speeding up the regulatory process, whose delays he said were unfair to the utilities and jeopardized their ability to provide adequate service. The administrative changes that caused the most controversy were designed, he said, to reduce regulatory lag. In December 1970 the commission announced the first of these changes, a new personnel policy under which staff members were to rotate periodically from one assignment within a division to another. For example, hydraulic engineers would be asked to work in the gas division or an energy expert might be transferred to communications. Vukasin stoutly defended the plan as one that was necessary to meet the variable requirements of rate cases and to broaden the general competency of the staff.

The proposal met a fire storm of opposition, however. Critics charged that Vukasin was trying to make it easier for utility representatives to win their cases for rate increases by destroying the staff's expertise. They pointed out that a major utility often had scores of employees to collect and analyze data in support of its case whereas the CPUC, the nation's largest state regulatory body, had less than a half-dozen experts within any division who knew a particular company well enough to probe deeply its case. Irate staff members took their case to the media and to the legislature, which scheduled hearings and ultimately introduced several bills to block the changes. Among other bills was one that proposed to reverse the 1969 law and relieve the commission president of his duties as chief administrative officer. Vukasin eventually abandoned the plan but did not do so until he had transferred nearly one third of the commission's 755 employees and earned its enmity.[8]

A second series of reforms sought to streamline the hearings process. Three important changes were proposed: first, to prohibit anyone who was not an attorney from testifying at a hearing on behalf of any group; second, to create a "limited status" category for groups without legal representation, such as consumer groups (a step that would deny such groups the opportunity to cross-examine witnesses, including company representatives); and third, to establish a "special interest" test to limit the number and kinds of groups that might appear at a commission hearing to those with a direct interest in the case at hand.[9] Vukasin maintained the reforms were needed to reduce the unnecessary delays in proceedings before the commission. The delays were due in large part, he felt, to uninformed and excessive participation by public witnesses. As an example of the problem, Vukasin cited a 1971 case that had had "eighty-one days of hearing, two hundred witnesses, unlimited cross-examination by anyone who walked in off the street and wanted to cross-examine, even though he had not heard the direct testimony and wasn't familiar with what had transpired before—eighty-one volumes, over a million and a half words of testimony in this case!"[10]

Reactions to Vukasin's Innovations. To Vukasin's critics and the press, the move was interpreted as just another attempt to spare the utilities criticism and to facilitate the approval of rate increases. The legislative hearings that were convened to hear opposition to the staff rotation plan also investigated the procedural proposals. Most consumer groups appear at hearings to cross-examine rather than to present testimony. Thus representatives of these groups argued that the creation of a limited status category would effectively shut them out of the regulatory process. After much debate the commission withdrew the proposals.

The staff's reactions to Vukasin's initiatives were uniformly negative. Hearings examiners were especially critical of procedural changes in the hearings process. One said, "When Vukasin started screwing around with the ground rules all hell broke loose, and this brought *us* into disrepute." The staff did not like the idea of consumer groups being kept out of the hearings process, either. One staff attorney said, "Historically, most intervenor groups have tracked the staff. Their witnesses will take a lower figure on the rate of return than the staff and this will make the staff look reasonable. Another advantage is that they can appeal a commission decision, and we can't."

If staff members were upset over the procedural changes, they were outraged over Vukasin's personnel orders, which they saw as a threat to their autonomy. Predictably, they fought back by trying to thwart the commission's plan of speeding up the regulatory process. Their methods of resistance consisted chiefly of ignoring or delaying the execution of directives and continuing to generate their own testimony as they saw fit. Staff attorneys directed to avoid taking independent positions in the briefs they filed in rate cases simply ignored the admonitions. Technical experts lengthened their investigations of utility cost surveys rather than shortening them. Relations between the staff and the commission became increasingly strained. Commissioners spoke of an "us-against-them attitude," and staff members charged that the commission was "looking for shortcuts." Instead of reducing regulatory lag, the commission's efforts increased it. The net effect was slipshod work: "You hold hearings for a year and then when they're drawing to a close, everything goes to hell. Decisions just get smashed together."[11]

Throughout Vukasin's term as president, nearly every one of the commission's efforts to reduce regulatory lag was greeted by this perverse slowdown response. At one point Vukasin confided to an interviewer that the staff had remained largely independent: "There is a firm condition over there that no commissioner ever tells staff members that they cannot introduce something they want to. We may ask them to do something else they had not intended to do, but we never tell them they can't."[12]

In large measure the staff's day-to-day routine of investigating complaints and handling inquiries was unchanged by the commission's managerial moves. Some disgruntled staff members who saw opportunities for advancement cut off by the new composition of the commission left the CPUC; others stayed, became demoralized, and hence became a source for journalists and consumer representatives.

Reducing Organizational Strain: The Fuel Cost Adjustment Clause

When the Reagan commission failed to reduce regulatory lag through a staff-oriented policy, it attempted to institute reforms that bypassed the staff. The most successful of these from the point of view of reducing regulatory lag was the fuel cost adjustment clause that allowed utilities to automatically raise rates to reflect increased energy costs without requiring them to go through lengthy rate cases first. This regulatory innovation was ultimately approved by nearly every commission in the country. Actually, in 1973 there was an energy cost adjustment clause on the books in California, but its importance to utilities became critical only after the oil price explosion. Sensing this the Vukasin Commission began to rely extensively on the fuel cost adjustment clause to assist utilities in raising rates rapidly. One of the commissioners even went so far as to call the fuel clause a "godsend."

This favorable view turned out to be vital to the utilities when the increases in the cost of fuel resulted in rate hikes that exceeded any in history. In 1973–1974 the CPUC authorized $608 million in pass-through rate increases to electric utilities without public hearings but authorized only $96 million in general rate increases that required public hearings and intensive scrutiny.[13]

In states that allowed a fuel cost adjustment clause, the measure was controversial, but in California in 1974 and 1975 the weather conspired to make it especially so. Historically, the energy adjustment clause had been based on a utility's fuel needs in an average year. If a utility relied on fuel oil to generate half of its output on average, its rates would be allowed to rise to reflect the cost of purchasing that much fuel oil. In dry years, when there was a shortage of water and hence of hydroelectric power, a system might have to substitute fuel-oil-generated power for hydroelectric power. The company would be able to raise rates based only on the historical average of fuel consumption, however, and thus would undercollect. On average the undercollections would be balanced out by overcollections. In a very wet year in which there were also sharp increases in the price of fuel oil, a utility might overcollect by such large amounts that it could take years before the formula resulted in compensating undercollections. In 1974 and 1975 in California, wet years coincided with the quadrupling of oil prices. The utilities collected hundreds of millions of dollars more than they paid out in costs. At least one utility, Southern California Gas, refused to lower rates in compensation. In this it was basically supported by the Reagan commission.

The overcollections were not lost on California consumer groups, however; they were just one additional source of controversy that served to bring the commission's behavior more and more into the public eye. By early 1974 rate increases in electricity and gas were coming almost monthly due to the fuel cost adjustment clause. It was in this setting that Pacific Gas and Electric Company sought a $223 million general rate increase—an increase larger than any in California's history.

Lifeline Rates: The Type III Case of Conflict Maximization

1974 was a propitious year for mounting a campaign to alter the structure of electric utility rates in California. Governor Reagan had announced that he would not seek another term in office, and a host of potential successors were eager to develop issues to run on. The controversy surrounding the CPUC and its approval of rate increases was a magnet to political entrepreneurs. Jerry Brown, his rivals, and half the California legislature eventually included sharply-worded attacks on the CPUC in their campaign rhetoric.

The Citizen Action League's "Electricity and Gas for the People"

But the established politicians were not the only ones seeking to exploit the CPUC's problems. The commission provided a valuable "enemy" around which consumer groups could develop organizational strategies and a stable incentive base. The most successful of these was a group that came to be known as the Citizen Action League (CAL). CAL's campaign against the PG & E rate increase, "Electricity and Gas for People" (E & GP), evolved into a campaign for lifeline rates. In the process of its campaign, CAL became one of the most important actors in the CPUC's political environment. Along with other consumer and environmental groups, CAL's presence brought about a major loosening in the external constraint faced by the commission.

The CAL campaign was born in January 1974 when a group of about one hundred people who had been active during the 1960's in civil rights and the anti-war movement met in the San Francisco Bay area at a conference sponsored by Organize, Inc., a training center dedicated to the ideas of the late Chicago activist Saul Alinsky. The conference had the purpose of "reviewing the lessons of social activism learned in the sixties and devising organizational strategies for the seventies."

The meeting was organized by Mike Miller, who had trained with Alinsky and had led a successful neighborhood group in San Francisco called the Mission Coalition. Miller was aided by Tim Sampson, an instructor in social work at San Francisco State College. A key purpose of the meeting was to identify an issue that would be useful in building a coalition of low- and middle-income people. The leaders wanted to build an organization that could eventually be used "to change the basis of power and work for economic justice."

CAL's Search for a Coalition-Building Issue.   A number of issues were considered at the conference. The OPEC price increases made the oil companies a natural target, but the leaders worried about how to organize an anti-oil-company campaign. The Pacific Gas and Electric Company seemed to be a good alternative. Tim Sampson recalled:

> *If we had had a handle on the oil companies, we would have gone after them. Instead we settled for the "mom and pop" store of the oil industry which was the electric and gas utility. Someone had brought to the meeting a clipping from the newspaper, announcing that PG & E was asking for the largest rate increase in history on top of huge rate increases for fuel offsets. We looked at the possibility of fighting PG & E as an organizing issue. It got us into the energy field which was good. It gave us a chance to fight a large corporation which was a monopoly. Plus it had its headquarters in San Francisco where most of our support was, as did the CPUC which regulated it. It just looked good all the way around.*

Shortly after the January conference a smaller group met to draft a platform for the E & GP campaign. Since the group's primary purpose was to build an organization by using the utility as an adversary to work against, its leaders were much more flexible on the definition of issues than they would have been if the campaign had been approached from a different perspective. As Mike Miller was later to write:[14]

> *The victory of the lifeline campaign cannot be understood other than in its context of building what "new populists" call majority constituency organizations. Our fundamental purpose in the campaign was not only to win a victory that would provide important benefits to low and mid-line income consumers, but to build an organization in which these consumers became active citizens. We see ourselves as organizers whose primary job is to build people power. The basic thing we wanted to do was build an organization that could challenge big business, make bureaucracy more accountable to the public, and force the political parties and elected officials to take stands on issues that we defined by organizing the people.*

Lifeline rates proved to be the issue on which the utilities, the CPUC, and the politicians were most vulnerable, but that was not immediately apparent to the CAL leaders. Sampson, who was the chairman of the E & GP campaign and later the president of CAL, said the group "initially zeroed in on stopping the rate hike." Blocking all rate increases seemed to be too difficult in the face of rapidly rising costs, however, so E & GP shifted its focus to the structure of rates:

> We actually picked lifeline to focus on after we were well into our utility campaign. From the beginning both lifeline and fair share rates were part of our platform, which also included stopping the rate hike, a fair chance for public power, and an end to price fixing and profiteering. Our focusing in on lifeline had an immediate organizing value. As we experimented with it, it was apparent that everyone was really responsive to the notion, particularly to the notion that lifeline would be for everybody.

The Lifeline Concept

Before tracing the campaign to adopt lifeline rates, it will be useful to digress briefly to describe the lifeline concept and to evaluate the claims put forward on its behalf by its proponents. In California, the concept of a lifeline rate for utility services was first expounded in 1967 by Edward L. Blincoe, president of the Utility Users League, in a case involving Pacific Telephone and Telegraph Company. Blincoe argued that the company should provide a special low-cost limited-service telephone rate for senior citizens and shut-ins so that such persons could check on each other and summon help in emergencies. The CPUC authorized the establishment of this special rate, which was called a lifeline rate, but made it available on an optional basis to any customer without regard to age, physical capacity, or income.

Since 1967 there have been repeated attempts to persuade the California commission and other state commissions to extend the lifeline concept of pricing to other utility services, notably electricity and gas. Proposals vary, but basically the lifeline concept is represented in any rate reform measure that gives low-volume users a respite from rising electricity bills by actually reducing their bills or by exempting them from particular rate increases. Most legislative proposals have set the lifeline quantity of electricity at 200 to 500 kilowatt-hours per month, although the figure has ranged to above 1500 kilowatt-hours in special circumstances. Proposals also vary in their applicability. Most apply to all residential consumers, but some would limit the lifeline rate to senior citizens or to persons who meet a "means test"—a device for confining the special rate to poor people with incomes below a specified level. The methods whereby revenues lost from the lifeline subsidy would be recovered also vary. Some proposals call for increased rates for industrial and commercial customers as well as for large residential customers, but others limit the offsetting rate increases to large residential users. A few make no mention of how the subsidy will be recovered.

Arguments For and Against Lifeline. Proponents of the lifeline concept advance three main arguments in its defense. The first is that access to electricity is a basic human right and that it is impossible to survive in a modern society without electricity. Lifeline rates would provide the minimum essential electricity requirements of the average user and thereby assure that no one would be impoverished by electricity bills. Critics respond that if access to any commodity is a basic human right, surely that commodity is food; yet, they assert, society does not give discounts for the first quart of milk or first dozen eggs any consumer buys each month. A related but different claim is that lifeline rates can be justified on equity grounds as an income redistribution measure. When presented as such, the concept has sometimes met with more approval, especially if proponents argue that its benefits should be restricted to a certain needy group. However, when proponents argue for a general lifeline rate for all residential users and justify it on the grounds that low-income persons are also low consumers of electricity, commissions tend to reject the view. In state after state it has been demonstrated that as a general income redistribution scheme lifeline is inefficient. Some poor persons who are high users receive little or no benefit under lifeline proposals whereas other persons, the rich who have second homes, may receive a benefit even though they do not need it.[15]

A second argument of lifeline proponents holds that lifeline rates, to the extent that they replace the declining block rate structure with flat or inverted rates, will encourage energy conservation. If people have to pay more rather than less for additional usage, they will have an incentive to cut their consumption. Critics suggest that the conservation effect of lifeline rates will depend upon the level chosen as a lifeline amount and on the way the revenue loss is recovered. The conservation effect will also depend upon how industrial users respond to the higher prices. Lifeline rates, if adopted for all residential users without regard to age or income, would confer a benefit not only on people who use less electricity than the lifeline amount but also on some people who use much more. That is because everyone would receive the first 250, say, kilowatt-hours at a low rate and then be charged a somewhat higher rate for additional consumption. At just what level the surcharge would equal the subsidy depends upon the specific recovery method adopted. And, in turn, the recovery method will likely affect how much conservation consumers can be expected to achieve.

Consider the following example. A lifeline subsidy of one cent per kilowatt-hour (kwh) is given on the first 250 kwh of use. The system sells 100 million kwh to customers using up to this amount. The revenue loss is therefore $1,000,000. If the cost of the subsidy is completely borne by the residential consumers using more than the lifeline amount and if sales to this class of users equal an additional 100 million kwh, the nonlifeline surcharge would equal one cent per kwh. Under this scheme, the residential consumer using 250 kwh per month saves $2.50. The breakeven point—the point at which a consumer is neither helped nor harmed by the new rate—is 500 kwh. At 500 kwh the lifeline benefit has been fully offset by the surcharge. The net change in residential bills is shown in Table 5.1.[16]

If the lifeline-induced revenue loss were recovered from among nonlifeline users of all classes, including industrial and commercial users, the effect of the lifeline rate would be quite different. If, in the above example, the nonlifeline usage equaled an additional 400 million kwh, the nonlifeline surcharge would be only 0.25 cents per kwh. Residential users who used less than 1250 kwh would actually see a reduction in their total monthly bill and might therefore increase, rather than reduce, their consumption of electricity. One

Table 5.1 The Impact of a Lifeline Subsidy on Electricity Rates (Surcharge Applied to Residential Users Only)

1. Lifeline Subsidy per kwh	1 cent
2. Kwh Billed Below 250 kwh	100 million kwh
3. Revenue Loss (1 × 2)	$1 million
4. Nonlifeline Sales, Residential Class	100 million kwh
5. Surcharge (3 ÷ 4)	1 cent
6. Net Change in Residential Bills	
250 kwh	−$2.50
500 kwh	0
750 kwh	$2.50
1000 kwh	$5.00
1500 kwh	$10.00
2000 kwh	$15.00

Table 5.2   The Impact of a Lifeline Subsidy on Electricity Rates (Surcharge Applied to All Classes of Users)

1. Lifeline Subsidy per kwh	1 cent
2. kwh Billed Below 250 kwh	100 million kwh
3. Revenue Loss (1 × 2)	$1 million
4. Nonlifeline Sales, All Classes	400 million kwh
5. Required Surcharge per kwh (3 ÷ 4)	0.25 cents
6. Net Change in Residential Bills	
250 kwh	−$2.50
500 kwh	−$1.88
750 kwh	−$1.25
1000 kwh	−$0.63
1250 kwh	0
1500 kwh	$0.63
2000 kwh	$1.88

certain effect of this system would be that the rates of industrial and commercial users would rise. This rate hike might have the effect of inducing such users to cut back on their consumption of electricity, but it might also have the effect of encouraging them to generate their own electricity, resulting in little or no overall conservation of fuel. (See Table 5.2 for this second case.)

The final argument for lifeline rates is that they are economically justified for a number of reasons—some of which are specious. In Chapter 4, it has already been pointed out that a lifeline-like result is not necessarily incompatible with an efficient rate structure. If the revenues generated by marginal-cost-based rates exceed those deemed reasonable for a utility to earn, if commissions decide to distribute those excess returns in accordance with the inverse elasticity rule, and if—to make this line of reasoning thoroughly convoluted—commissions determine that the demand for electricity in the first 250-kwh (or so) block of use is the most inelastic, then a lifeline-like rate might be economically justified. Efficient pricing does not require lifeline rates, however. It is only when the above three conditions are met that the two concepts are compatible.

In California, the inverse elasticity rationalization was not the primary economic argument advanced by consumer groups on behalf of inverted and lifeline rates. Instead, their argument was built on the following assumptions. First, rates are rising because utilities must build expensive new plant capacity to meet growing demand. High-volume users are responsible for this added demand; therefore, they should pay higher, rather than lower, rates for each additional block of electricity they use to reflect the true costs the system incurs to serve them. Two, most utility systems rely on a mix of electricity sources: hydroelectric generators, gas or coal-fired turbines, and some nuclear generators. Of these sources, hydroelectric generators are by far the cheapest to operate. If these cheap power sources were reserved for residential use, lifeline quantities of electricity could be made available without subsidy.[17]

Other Arguments Against Lifeline. To an economist, these arguments are wrong. An economically efficient price just covers the cost—private as well as social—of producing the last unit of output. An alternative way of saying the same thing is that price should equal the amount of money that could be saved if the consumer decided to do without the additional consumption. If a system can save the same amount when a small user reduces its output by, say, 100 kilowatt-hours per month as when a large user reduces its output by an identical amount, the large user ought not be required to pay more. For the same reason, there is no economic justification for segmenting the supply curve by allocating power produced from low-cost sources to residential use. If residential users cut back their consumption, the utility would save not the cost of producing hydro power but rather that of producing the more expensive turbine-generated power. Thus residential consumers ought to pay the full cost of those potential savings. To argue that they should only pay the cost of producing hydroelectric power is to argue that it was only the power demands of industrial and commercial users that required the utility to expand its capacity to include high-cost sources of power. Further, it is to argue that cutting back on residential demand will not economize on these new sources. Both of these arguments neglect the fact that it is extra demand taken during the peak period that results in costly new additions to plant capacity. High-volume use, if taken off-peak, may actually reduce costs.

The most important critics of lifeline rates from a political standpoint are not economists, of course, but industrial and commercial users, who believe that the lifeline proposal is nothing more than a scheme to redistribute the burden of paying for electricity from residential users to them and the electric utilities. At first, one might wonder why utilities whose overall revenues will not be reduced by lifeline should care about how rates are structured. Utilities argue that the declining block structure is important in preserving the stability of revenues. They assert that when demand for electricity varies, overall revenues are affected less when the rate for high-volume usage is low rather than high. Lifeline proponents maintain that the revenue stability argument is only a smoke screen for the companies' real desire to grow. Since the declining block structure promotes growth, any departure from that pricing method jeopardizes management desires to expand.

CAL's Lifeline Campaign: The Formation of a Successful Coalition

There are four major reasons rate structure reform proved to be a useful organizing issue for the Citizens Action League, and these reasons in turn help explain why the proposal for lifeline rates—not some other proposal— was adopted in California.

The Poor and the Elderly. First, the rapid increases in electric and gas rates had hit the poor and the elderly especially hard. A campaign to relieve them from the burden of rate hikes was attractive not only to groups acting for the poor and the elderly but to many other groups as well. CAL chose to fight for universal lifeline rates that would apply to all residential consumers, however, their position being that people were tired of special needs programs financed by the middle class. This stance appealed to consumer groups such as TURN (Toward Utility Rate Normalization), which was already developing testimony to be entered in the PG & E case and which was opposed to any rate structure reform that failed to benefit all residential consumers. The non-exclusive lifeline proposal appealed to senior citizen groups as well. In California these groups had won a number of legislative victories, but some of them feared that their victories tended to isolate them from other consumer organizations; others objected to special needs programs because they felt them to be stigmatic and degrading. CAL's position was attractive because it offered senior citizens a direct economic benefit and at the same time opened up a political opportunity to work with other groups.

Labor Unions. Rate structure reform also appealed to organized labor, another key constituency in the "Electricity and Gas for the People" coalition. In organizing the E & GP campaign, Tim Sampson said CAL "tried to be careful that no other oxes were gored." Labor, which proved to be dividable in other states, worked solidly for lifeline in California. Of key importance to the unions was the issue of jobs. E & GP sought to avoid taking an anti-growth position, which is anathema to the unions. One of the unions' main reasons for shifting away from active opposition to the rate hike was that E & GP made the credible threat to union employees that if a rate increase was not obtained jobs would be cut and wages frozen. By focusing on the reform of the rate structure, E & GP was able to defuse the job threat. The extra revenues would still be there for wage increases; they would simply be paid for by a different class of consumers. Industrial users have turned this argument around in other states that have considered lifeline, saying that if their energy bill goes up they will be forced to relocate outside the state. In Maryland and Massachusetts large industrial users have organized their employees to oppose lifeline legislation, using the slogan "if you don't have a job, your utility bill won't matter." In California the industrial users failed to mobilize their employees, however.

Environmental Groups.   A third important reason CAL focused on the rate structure was that such a focus appealed to environmental groups, several of which were already at work in other states to bring about an end to the declining block structure. In California, the Planning and Conservation League (PCL) and the Environmental Defense Fund (EDF) were blaming the declining block structure for promoting unneeded power plant construction. These costly new facilities were a significant causal factor behind the rate increases, they said. Initially, the environmentalists seemed to provide E & GP with useful allies in the fight against traditional ratemaking. Later, as lifeline rates became the focus of the campaign rather than the environmentalist-preferred concept of peak load pricing, the environmentalists' support waned. The EDF, whose members were predominantly middle and upper middle class and which was more interested in conservation and environmental preservation than in low utility rates, dropped out of the coalition in dissatisfaction over TURN's lifeline witness in the PG & E case, Eugene Coyle. The group sponsored its own expert witnesses, two economists, Charles Cicchetti and William Vickrey, whose pricing testimony in the PG & E hearing was based on principles of marginal cost. The Planning and Conservation League, meanwhile, came under pressure from some of its board members to abandon lifeline when it became clear that the chief conservation issue in electricity was not just how much electricity was consumed but at what time of day it was consumed. Some PCL board members joined the critics of lifeline who argued that subsidizing the lifeline block might have the effect of encouraging consumption rather than conservation. These internal disputes resulted in the PCL becoming less active as the campaign progressed, but it never openly attacked lifeline. On occasion the league even provided lobbying support through its Sacramento office.

Rate Reform as a Viable Issue. Finally, CAL's focus on rate structures provided a positive alternative to the rate hikes which seemed viable. Mass citizen action groups like E & GP often frame issues in negative terms because less power is required to block or attack programs than to initiate their own programs. In the case of the E & GP campaign, however, it would have been impossible to block entirely the approval of the PG & E rate increase. Even more than rate increases public utility commissioners fear service failures. If the utility could not raise the capital it needed to meet operating expenses, it might be in jeopardy of bankruptcy and of service inter-ruptions, brownouts, or service failures, blackouts. To ask the CPUC to risk such occurrences would simply have not been real-istic, and to hope that the Reagan commission could be persuaded to deny the increase would have been foolish. Furthermore, even if the CPUC did authorize a rate increase of as much as 25 per cent below the requested amount, E & GP could hardly claim a victory because the CPUC regularly approved less than the sought-after amount.

A campaign for lifeline however, appeared to be something a mass citizen action movement could win, and it was a win the movement could claim credit for. The lifeline concept had never been tried on California electric rates. The utility could not threaten bankruptcy because it would still get its required revenues. Exactly who would pay the extra costs could be left unsaid for the moment. It seemed reasonably clear that it would not be the residential consumer, and that was what mattered to the consumer groups. Moreover, the proposal had wide appeal to politically active consumer groups, and it promised an immediate and visible impact: no more rate increases for the lifeline amount of electricity and gas.

CAL's Action Strategy

Having defined an issue that would be useful in building their or-ganization, E & GP leaders next sought an action strategy. Ulti-mately, it would be two years before the PG & E rate request was decided by the commission. CAL knew on the basis of previous rate hearings before the CPUC that it would take eighteen months, at least, before a decision on lifeline would be reached. This both served and hindered CAL's strategy. A long campaign was a drain on resources, but at the same time the unresolved rate case was CAL's best source of incentives for organizational development. Then, too, since the CPUC was undergoing a change in membership, what had started as an external fight for lifeline might—and in fact did—become an internal fight at the CPUC.

On February 5, 1974, less than a month after CAL's January organizing conference, over four hundred people converged on a CPUC hearing dealing with the proposed $233 million rate increase. With them came the news media. After delivering a brief statement attacking PG & E and the CPUC's failure to regulate it, the group left the CPUC headquarters to march about one mile to PG & E's corporate headquarters on Market Street. The news media followed them to hear a company public relations spokesman refuse to accept E & GP's demands that the rate request be withdrawn and that a new request including lifeline rates be drafted. Finally, company security guards locked the building to keep the E & GP crowd out of doors. The evening news reported all these events in great detail, and the E & GP campaign was launched.[18]

CAL campaign leaders settled on three strategies: (1) confrontations with PG & E; (2) direct action aimed at the CPUC; and (3) participation in the electoral process. A fourth strategy, intervention with expert witnesses in CPUC proceedings, was eschewed by E & GP but employed, as mentioned above, by two other early members of the CAL coalition, TURN and the EDF. CAL sought confrontations with PG & E to generate publicity and to expose the corporation's salary structure and profits. It attacked the CPUC as being the rubber stamp of the utility industry and as being closed and secretive. The group interrupted commission hearings and entered closed meetings, demanding that the meetings be opened to the media and the public.

All of these efforts provided useful incentives to the CAL organization. Watergate had heightened the press's interest in secretive government deliberations, so CAL's charges received a good deal of coverage. After the CAL organization succeeded in opening up the CPUC conferences, commission staff members scoffed that all the consumer advocates had accomplished was to provide utility representatives with more information on how the commission reached its decisions, providing no similar benefit to consumers since the groups had stopped coming to CPUC conferences after they had won the right to do so. What such statements overlook is the value of successfully "opening up the process" in generating purposive incentives for the CAL organizers.[19]

The E & GP leaders recognized from the start that they could not win lifeline rates through the CPUC's normal decision-making process as long as Reagan appointees controlled the commission's board. The majority could be counted on to oppose the measure steadfastly. After its initial confrontations with the CPUC and PG & E, the group concentrated on making the issue of rate reform an important part of the 1974 legislative and gubernatorial campaigns. In this they were not so much forcing the issue as acting in concert with the politicians, who had not failed to recognize the value of utility rate reform as a campaign issue. Nearly every gubernatorial candidate promised to make the CPUC a more responsive and open body.

CAL's Search for Legislative Support

Throughout the summer and fall of 1974, E & GP sought pledges of support for a legislatively mandated lifeline rate from candidates for the legislature. An early and important ally was the Speaker of the California Assembly, Leo McCarthy, who next to the governor was probably the most powerful elected official in Sacramento. The Speaker determines, among other things, which committee will hear a bill in the assembly. Later, favorable assignment was very important to the lifeline bill's success. McCarthy wrote Mike Miller on September 20, 1974, to indicate his support for the lifeline concept in PG & E rate schedules.[20] Privately, he worked with E & GP on a draft of a bill he had agreed to introduce in the assembly.

After McCarthy's public endorsement of lifeline, PG & E representatives began meeting with E & GP to discuss rate revision— something they had refused to do before. In November, PG & E general counsel Frederick T. Searls and associate general counsel Malcolm Furbush met with some 400 E & GP members to urge the group to work for an energy stamp alternative to lifeline. Under this alternative, energy stamps, like food stamps, would be given to a select, low income population that met a means test. The advantage of energy stamps over lifeline rates, the PG & E representative said, was that a more meaningful level of assistance could be given to people who really needed it. There was a delightful irony in the energy stamp proposal which was not missed by the media: A large business corporation was advocating a fairly traditional welfare measure, and a consumer group made up of poor people and senior citizens was objecting to it on the grounds that "it would create a huge bureaucracy and would be expensive to administer."[21] E & GP's opposition was an important reason the energy stamp proposal was not accepted by the governor or the legislature when the utility and the industry later made a concerted push to get the proposal adopted in place of lifeline rates.

The PG & E representatives made little headway with CAL in November, but E & GP claimed to have won a major victory: it had been recognized by PG & E as a legitimate organ of consumer sentiment. Miller stressed the importance of such confrontations throughout the campaign, calling them "win-win" situations: The CAL campaign won if PG & E refused to meet with it because the refusal proved PG & E's arrogance; and it also won if PG & E agreed to meet with it because the agreement was a tacit recognition of CAL's strength.[22]

Although CAL was lining up sponsors for its lifeline bill in the fall of 1974, it did not have to wait as long as it had expected to before a lifeline bill was introduced in the assembly. Unbeknownst to E & GP leaders, Assemblyman John Miller of Berkeley had taken a copy of CAL's lifeline platform, made it into a bill, and introduced it into the assembly on his own on December 4, 1974.

The CPUC Lifeline Decision

The legislative history of the California lifeline bill is laced with the fascinating stories of intrigue, compromise, and coalition building which make up the history of any important piece of legislation. While the details of this history are interesting, they are not germane. Of greater interest is the California Public Utility Commission's response to the political battle for lifeline. When the PG & E case was decided on September 16, 1975, the commission ordered the utility to adopt lifeline rates. That decision was issued one week before the governor signed the legislature's bill into law. Informed sources maintain that without CPUC support the lifeline bill could not have passed the legislature. That support has to be explained.

CPUC Opposition to Lifeline

As late as January 1975, the CPUC—commissioners and staff—was solidly opposed to the lifeline concept. Commissioner Vernon L. Sturgeon, who had replaced Vukasin as president of the CPUC, said bluntly:

> *Lifeline is a fraud. No one gets lifeline because it causes higher rates for business and they pass on more than that cost of business to consumers. It's good for PR and nothing else. They tried to push it as conservation, but that's pure unadulterated b___ s___.*

After lifeline was adopted, Commissioner William Symons, Jr., Reagan's first appointee, said, "We've become a welfare agency— giving it to people who don't deserve it. We've gotten completely away from the cost-of-service idea. Now it's just like throwing darts at the wall."

Most members of the CPUC staff agreed with the Reagan commissioners on lifeline. A senior staff examiner confided that in his opinion lifeline was "misguided compassion. All the examiners who have worked on it think it is wrong." The opposition extended to the utility division, where one engineer voiced the opinion of many, "Lifeline is an attempt to sell a 'free lunch' and there ain't no free lunch." On February 25, 1975 the commission secretary wrote Assemblyman John Miller to express the CPUC's opposition to Miller's proposed bill.[23]

In March, the CPUC attorney assigned as counsel to the PG & E proceeding filed the staff's brief in that case. The brief took a negative position on the lifeline testimony entered in the case by TURN's witness, Eugene Coyle, during the hearings in November 1974:[24]

> *This is an appropriate point to discuss the lifeline concept. In doing so it is well to remember Vickrey's generalization "that a rate structure that comes closer to reflecting the relevant cost elements will be fundamentally more equitable than one which distorts those cost elements."*
>
> *The distortion of the cost elements by TURN's rate is extreme, and the degree of subsidy to the domestic class, which would result from TURN's proposal, is greatly increased.*
>
> *Such dramatic changes in rates to the domestic class are advocated by TURN despite a dearth of evidence that the overwhelming majority of domestic customers are in need of significantly higher subsidies. . . .*
>
> *Some other mechanism for relieving the deserving poor from high electric rates is certainly indicated. Already employed by governments within this state are individuals whose job it is to determine eligibility for welfare purposes. Relatively minor changes in existing laws, to be*

*effected by the legislature, could activate this system to relieve the
deserving poor of additional burdens created by utility rates. Spread-
ing the cost of this additional social welfare responsibility over the
taxpayers as a whole would certainly make more sense than assigning
the burden to the large domestic customers (some of whom may be
equally deserving of government support), or the commercial customers.*

The Appointment of Leonard Ross to the CPUC

The CPUC's solid wall of opposition to lifeline began to crumble in
February 1975 when newly elected Governor Brown made his first
appointment to the commission. Brown had promised to appoint
commissioners who would open up the process of regulation and be
more accountable to public opinion. His first choice for the job was
Leonard Ross.

Ross's credentials for the position of commissioner were impressive. He had earned degrees in law and economics from Yale and,
with James Tobin, had authored a number of articles on inflation and
economic growth. In addition, he was the co-author, with Peter
Passell, of the book *The Retreat From Riches*, a liberal argument for
economic growth. He had been an assistant professor at the Columbia University School of Law from 1971 to 1973; chairman of Mayor
Bradley's Cost-of-Living Council in Los Angeles in 1974; a member
of the Mayor's Panel on Public Utilities in New York City in 1969;
and Governor Brown's transition budget advisor in 1974—all before
reaching the age of thirty. In fact, at the commission, Ross was soon
referred to as the "whiz kid." (As a child Ross had actually been a big
winner on the $64,000 Question, a TV game show.)[25]

From the start Ross made energy rate structure reform one of his
three main objectives, along with "opening up the process" and
"maximizing the degree of competition in California's trucking industry." He was knowledgeable in the economic theory of regulation
but recognized "practical problems" in administering efficient
rates.[26] Some of Ross's acquaintances said that he was politically
ambitious and that he aspired to be California's treasurer and, ultimately, a United States Senator. Of his own plans, Ross would only
say that he "did not envision a life as a regulator here [in California]
or on the federal bench" and that he would not "work for the utilities
later, either."

Whatever his motivations, Ross soon became a vigorous proponent of lifeline rates. On March 6, 1975, he directed the commission secretary to write Assemblyman Miller to inform him that he, Ross, had not been appointed to the CPUC when the commission had decided to oppose the bill.[27] In the same month he further indicated his support for the lifeline concept by choosing Jim Cherry as his legal assistant. Cherry, a close friend of Tim Sampson of E & GP, had represented consumer groups in favor of lifeline at the commission hearings on the PG & E case.[28] With Cherry, Ross worked out a strategy to "sell" lifeline. The two decided to define access to electricity as a 'basic human right' and to portray lifeline rates as an aid to conservation rather than as an income redistribution measure. Said Cherry:

> We never from the start said it was a perfect match on income redistribution grounds. It was clear where utility bills were headed and who was paying the most and that the rate structure was exacerbating the problem . . . I kept trying to get the conservation groups to come with us and to see that if you want inverted rates this is the only way to get it.
>
> What I was dead set against was another nice program for poor black folk. My cut was, "it's you and me." If you're a small residential consumer, you ought to get your pot sweetened. And you ought to do something to stop getting whacked by rate increases.
>
> Lifeline is not just a small cookie that nice white folks are going to give to black people. This is not a poor person's bill. It's a bill for you and me.
>
> People who kept saying, "but lifeline doesn't help the poor," just didn't understand the issue. The issue is: what is the basic amount society can afford to give you and me? I'd keep explaining that, but they'd come right back and ask what it did for the poor. They didn't understand the broad based political support for lifeline without restrictions on income.

The Ross-Batinovich Compromise

In promoting lifeline rates, Ross's first target was Governor Brown's next appointee to the public utilities commission, Robert Batinovich. When Batinovich was appointed in March 1975, he was chiefly known to the public as an "enemy" of Richard Nixon, having earned his place on the former President's enemies list by contributing heavily to the political campaigns of George McGovern, Paul McCloskey, and Jerry Brown.[29]

Batinovich had some initial reservations about lowering the rate on the first block of consumption from its then prevailing rate which was the proposal in the bill before the legislature at the time—but he was receptive to the notion of doing away with the declining block rate structure. Batinovich and Ross hit upon a compromise that allowed a lifeline rate prospectively: The rates for the lifeline quantity of electricity would remain frozen at the current level, and further rate increases would be allocated to the utility's tail blocks until the rate structure was inverted.

Once the compromise was reached, Ross and Batinovich spoke out publicly in favor of a lifeline rate. On July 31, 1975, the two commissioners sent a letter to state Senator Alfred E. Alquist, chairman of the Senate committee that was supporting the Miller-Warren Energy Lifeline Act, the Public Utilities, Transit, and Energy Committee, encouraging the committee to adopt the compromise they had drafted.[30] On the same day Ross wrote Tim Sampson, thanking him "for the energetic and highly effective effort your organization has devoted to the lifeline campaign."[31] The Alquist committee was the last major hurdle before the bill came to the floor of the senate. The Ross-Batinovich letter and the commissioners' phone calls of support elicited the margin of support the bill needed to clear the committee.

The CPUC legislative liaison, Larry Garcia, whose job it was to represent the commission before the legislature, was placed in a difficult position. While Ross and Batinovich were lobbying the Alquist committee in favor of lifeline, it was Garcia's job to appear before the committee in opposition to the bill under direction of the commission's majority. Within the month that situation changed, however: The new commission president, David W. Holmes, a Reagan appointee, voted to support the lifeline position of Ross and Batinovich. On August 25, the CPUC secretary again wrote Assemblyman Miller, this time to indicate commission support for the lifeline bill as it had been amended in the Alquist committee.[32] On September 4, 1975, the bill passed the senate and went to the governor for his signature. The bill left it to the CPUC to designate "a lifeline quantity of electricity which is necessary to supply the minimum energy needs of the average residential user for . . . space and water heating, lighting, cooking and food refrigerating. . . ."[33] The CPUC was to take into account "differentials in energy needs caused by geographic differences, by differences in severity of climate, and by season."[34] Under the bill, the lifeline rate was to remain frozen until the average rate charged all customers increased 25 per cent or more. Even then the CPUC was not required to increase the rate on the lifeline amount.

The fact that Commissioner Holmes had agreed to support Ross and Batinovich on lifeline meant that the commission could issue its own decision in favor of lifeline in advance of the effective date of the legislation, an option which the CPUC exercised in the PG & E case that had stirred up the initial controversy. Later Ross would say that despite his commitment to rate reform he had not expected the turnabout in the commission's position to come as easily as it did: "No one expected Holmes to do what he did," he said.

In attempting to explain Holmes' vote, another commissioner, a Reagan appointee, said:

> *I'll be blunt. Holmes did a flip flop. . . . He heard we were going to make Sturgeon president again and that made him mad. So he went with them [the Brown appointees]. He wanted to be president [of the commission] so bad. He'd say "I've been president of every organization I've ever served on." There was some talk about Brown's reappointing him, but that was unrealistic.*

Holmes himself defended his votes, stating that they were consistent with his prior behavior on the commission and that they were evidence of his personal independence, "I've never sided with anyone, with neither the Reagan nor Brown people. I don't give a damn what they think. On important issues the Brown people are joining me, I'm not joining them."

The CPUC Staff's Response to the Lifeline Decision

The normal procedure in a rate case is for the assigned hearing examiner to summarize the arguments on the key issues and then recommend a decision. The commissioner assigned to the case may adopt, reject, or modify the examiner's opinion before circulating a personal proposal among commission colleagues in search of the needed two additional votes. The hearings examiner in the PG & E case, Parke Boneysteele, privately expressed his opposition to the lifeline concept and opted to avoid making a recommendation on the issue:

> *There was a lot of pressure to get the decision out. There were political decisions here. I met with Holmes and Ross at the scheduling conference and they asked how we could speed up the process. I suggested*

that we adopt a division of labor: I would concentrate on the rate of
return phase of the rate case and some of the staff working with Ross
and Holmes should do the rate spread [structure the rates]. I was more
or less the reporter when it came to what their wishes were in explain-
ing and rationalizing the decision. They never asked my view and I
didn't give it.

While Boneysteele adopted a somewhat philosophical attitude toward the lifeline decision ("when you're a bureaucrat you do what the political appointees want"), many of his fellow staff members were not so resigned. Most of the engineers in the utility division objected strenuously that lifeline would have a "distorting effect on cost of service." Attorneys who had followed the case found the decision objectionable because "the commission's adoption of rate design favorable to TURN was a result of political pressures rather than the record." One attorney, expressing a representative view, said, "I get terribly discouraged when the record is not considered; when a commissioner only takes his preconceived notions into effect and says 'if we're wrong let the courts show us.' I've heard it from the Reagan commissioners and I've heard it from the Brown folks too."

A different opinion was voiced by Tim Brick of the Los Angeles consumer group CAUSE, who termed the Ross and Batinovich style "regulation by press release": "Ross and Batinovich saw themselves as activists wanting to shake up an entrenched bureaucracy. I think they both play the media in the same way we do—regulation by press release, and I don't object to that."

Implementing the Lifeline Law

Despite the overwhelming staff opposition to lifeline, the CPUC's decision was relatively easy to implement. Hearings were held to determine lifeline quantities of gas and electricity in response to the legislative mandate. The task of preparing estimates of the minimum needs of the average residential consumer by climatic zone and by end use required some judgment, but after the decision was made little coordination was needed to implement the rate. The basic lifeline quantity for lighting, cooking, and food refrigeration was set at 240 kwh per month. An additional 250 kwh per month was allowed for consumers with electric water heating, and up to 1420 kwh per month more was allowed for residential users who relied on electric space heating in the mountainous areas of the state. Slightly

Table 5.3   A Comparison of Pacific Gas and Electric Company Bills, 1/1/76 and
            1/1/77

kwh/mo.	Bill at 1/1/76 Rates	Bill at 1/1/77 Rates[1]	Bill at 1/1/77 Rates Without Lifeline[2]
240	$ 7.73	$ 7.63	$ 8.61
500	14.16	15.17	15.99
1000	26.33	29.48	30.00

Source: California Public Utilities Commission.
[1] Basic lifeline allowance = 240 kwh/month.
[2] Theoretical calculation based on the assumption that increased charges would have been spread evenly among all sales levels.

lower allowances were approved for residential consumers living in master-metered apartments.

In February 1977 the CPUC filed a report with the legislature on the effects of the lifeline rates on California customers. While the price of the lifeline quantity of electricity was unchanged, PG & E's average charge on all other sales had risen 16.77 per cent. (A later report showed that some PG & E industrial rates had increased as much as 94.8 per cent.[35]) Table 5.3 summarizes the report's survey of PG & E rates.

A PG & E customer who used the basic lifeline allotment of 240 kwh per month realized a benefit of $0.98 per month as a result of lifeline rates. At a usage of 100 kwh/month the benefit was $0.41; at 200 kwh/month, $0.82; at 300 kwh/month, $0.94; at 400 kwh/month, $0.88; and at 500 kwh/month, $0.82.

With a lifeline subsidy of $0.0041 per kilowatt-hour and a surcharge of $0.0006 for usage in excess of 240 kwh/month, the break-even point was only reached at 1840 kilowatt-hours per month. Above that level, customers were billed more for electricity than they would have been billed for without lifeline. In 1976, 97 per cent of PG & E's residential users had bill reductions as a result of lifeline. The lifeline mandate had resulted in redistributing the burden of rate increases from residential users to commercial and industrial users, which was exactly what its proponents had hoped it would do.

Endnotes

1. *Wall Street Journal*, January 28, 1978, p. 14.
2. Public Utilities Commission of the State of California (hereafter referred to as CPUC), Application of Pacific Gas and Electric Company, Decision No. 84902

(September 16, 1975). Also, Assembly Bill No. 167, the Miller-Warren Energy Lifeline Act, Chapter 1010, Section 739 of the Public Utilities Code (September 23, 1975).

3. Interview by Douglas D. Anderson (August 1976). The chief of the utilities division at the CPUC, Walter Cavagnaro, added the following comment, "The quote on Berkeley engineers is of interest—we also have had some of the best from Stanford, Santa Clara, and many other California and east of California universities." Letter from Cavagnaro to the author (January 19, 1978).

4. *Pacific Telephone and Telegraph v. CPUC*, 62 Cal. 2d 634, (1965).

5. *Los Angeles Times*, December 22, 1974, Part I, pp. 1, 3, 27–28.

6. *Oakland Tribune*, February 18, 1974, pp. 33, 37.

7. The change required four of the commission's six division heads (utilities, legal, examiner, and administrative) to report directly to the president and to serve at the commission's pleasure. Previously, the division chiefs had enjoyed civil service status.

8. Although rotation had been practiced in all previous commissions, it had been done in an orderly manner. The Vukasin approach was wholesale and did not provide for an orderly transition period.

9. *California Journal*, April 1971, p. 105.

10. Ibid., p. 107.

11. *Los Angeles Times*, December 22, 1974, Part I, p. 28.

12. *California Journal*, April 1971, p. 105.

13. *Los Angeles Times*, December 22, 1974, Part II, p. 1.

14. Mike Miller, "The Electricity and Gas for People Campaign, 'E & GP': An Analytic History," unpublished paper (San Francisco: Citizens Action League), p. 6.

15. Joe D. Pace, "Lifeline Rates: Will They Do The Job?" *Public Power* (November–December, 1975), pp. 21–30.

16. A similar table appears in ibid.

17. See Eugene P. Coyle, "Rate Design Proposal for Pacific Gas and Electric Company," testimony entered in CPUC Application No. 54279 (November 18, 1974).

18. *San Francisco Examiner*, February 3, 1974.

19. See James Q. Wilson, *Political Organizations* (New York: Basic Books, 1973) for an analysis of the incentive systems of voluntary organizations.

20. Letter on file with the author.

21. *San Francisco Examiner*, November 3, 1974, Section A, p. 2.

22. Miller, "Electric and Gas for People Campaign."

23. Letter on file at the CPUC.

24. CPUC Staff Brief on Application No. 54279, (March 1975).

25. Peter Passell and Leonard Ross, *The Retreat From Riches* (New York: Viking Press, 1973).

26. Transcript of Hearings in CPUC Case No. 9804 (April 10, 1975), pp. 351–358.

27. Letter on file at the CPUC.

28. Transcript of Hearings in CPUC Application No. 54279 (February 4, 1974), pp. 23–27.

29. *Los Angeles Times*, March 27, 1975.
30. Letter on file at the CPUC.
31. Letter on file with the author.
32. Letter on file at the CPUC.
33. Assembly Bill No. 167, the Miller-Warren Energy Lifeline Act, Chapter 1010, Section 739 of the Public Utilities Code (September 23, 1975).
34. Ibid.
35. A copy of this report is on file at the CPUC. The report is comprised of data that was prepared by the utility for CPUC President Robert Batinovich. It was given to the author by Batinovich.

Chapter 6

REGULATORS, POLITICS, AND ELECTRIC UTILITIES

Among the conclusions to be drawn from the experience of the California Public Utilities Commission in reforming electric rate structures is this: There are times—even late in an agency's history —when regulators can formulate and implement important policies despite hostility and strenuous objection by client industries and their largest customers. As innocuous as this statement sounds, it is made interesting by the very fact that natural life cycle and related theories of regulatory behavior would lead us to expect just the opposite.

Another conclusion to be taken from the California experience is that it is possible to understand the circumstances under which one can and cannot deduce regulatory behavior from knowledge about the goals and preferences of an agency's executives. Let us take a closer look at this latter conclusion.

In the case of California we saw that when the composition of the CPUC changed under Governor Jerry Brown, a pro-lifeline decision was reached with alacrity despite unanimous opposition to the lifeline concept by an earlier Reagan-appointed commission. What is more, despite solid opposition by the staff of the CPUC, lifeline rates were easily implemented. Within a year of the effective date of the new rates, nearly all of PG & E's residential customers had benefited from them, which is precisely the effect intended by the rate's supporters.

Explaining Lifeline's Success in California

There are good reasons why lifeline succeeded despite the opposition it encountered from industry and the CPUC staff. First, the
regulated utilities were not the only, or the most important, organized interest groups to bring demands and offer support to the
CPUC. Indeed, as a result of the controversy and adverse publicity
generated by the pro-business Reagan commission and the quadrupling of oil prices, the public had become deeply suspicious of the
relationship between the utilities and the CPUC by 1974. Sensing
this, two political entrepreneurs, Mike Miller and Tim Sampson,
skillfully identified the lifeline issue and used it to develop a mass-
based consumer group. This provided the agency with alternative
political and polemical resources later when its composition changed
and it began to look for support from non-business sources. On the
general matter of rate reform, two other groups, TURN and the
Environmental Defense Fund, contributed added information resources (of varying quality) through the agency's formal hearings
process. TURN's witness, Eugene Coyle, was essentially discredited
in this process, but the group continued to be important in keeping
the issue before the media and in staking out an extreme position in
favor of a universal lifeline rate. By contrast, the EDF's witnesses,
William Vickrey and Charles Cicchetti, provided the commission
with impressive testimony on the value and cost justification of a
peak load rate, which helped to persuade the CPUC to include these
rates in its final PG & E order. This testimony also had the effect of
causing the EDF and the Planning and Conservation League to
reassess their own support for lifeline. Withdrawal of environmental
support may have slowed the drive for lifeline's adoption, but it did
not stop it. Because the EDF and the Planning and Conservation
League took seriously Jim Cherry's warning that opposition to lifeline
would jeopardize all efforts to reform rate structures, they never
openly attacked the concept. These developments essentially left
Leonard Ross and Robert Batinovich free to ignore business and
environmentalist opposition to lifeline. Eager to establish reputations for themselves as regulators who were responsive to consumer
interests, Ross and Batinovich had no difficulty putting their own
imprimatur on the lifeline issue.

Second, the nature of the lifeline decision itself, is important in understanding the commissioners's ability to achieve their ends with respect to the issue, despite nearly unanimous opposition within the CPUC's powerful bureaucracy. Lifeline had the characteristics of a "point decision" that could be stated in operational terms: Freeze the rate charged for the first 240 kwh of electricity per month until all other rates have risen twenty-five per cent or more. It was a policy objective that could be compared unambiguously with reality. Lifeline rates were either frozen, or they were not. Industrial rates either increased, or they did not. Furthermore, once the decision was reached, little coordination was needed to implement it. The utilities were given an order that was enforceable, and there was little or nothing any staff critic could do to stop or hinder its implementation.

In short, the lifeline decision was the easiest of all decisions to manage. The nature of the decision allowed Ross and Batinovich to ignore bureaucratic constraints; the nature of the political context allowed them to ignore pressures by the utilities and their largest customers. This experience suggests that whenever programs or goals require little coordination to implement and can be stated in operational terms, and whenever external constraints on regulatory behavior are lax, it is sufficient to know the objectives of regulatory executives and the "market" to which they make their appeal in order to predict and explain agency performance.

These conditions obviously will not always characterize regulatory situations. Tasks are often more complicated and external constraints more binding. In our first California case, for example, John Vukasin wanted to reduce regulatory lag in response to warnings by utilities that they were facing financial insolvency. In attempting to find a way to streamline the hearings process, Vukasin adopted a policy of rotating staff assignments and other procedural changes. He soon found, however, that all of his policy initiatives had backfired. Because of internal opposition evoked by the moves, Vukasin's efforts increased regulatory lag, rather than reducing it, and brought the commission under heightened scrutiny from outside the agency as well. In this case, staff opposition alone was sufficient to frustrate the designs of both the agency's executive and its regulated industry.

The chief difference between these two cases is the presence or absence of a bureaucratic problem—the need to induce cooperative behavior in order to achieve policy objectives. Vukasin failed to achieve his ends and ultimately had to change his policies because he failed to realize that the nature of the task at hand, reducing regulatory lag, required the cooperation of the staff. Ross, whose behavior was no less offensive to the staff, was more successful in implementing lifeline not because he was a better manager but because the lifeline policy did not require management.

Why Lifeline Failed in New York

Helpful as task analysis is in explaining the performance of the California Commission over time, it does not assist us in understanding why New York, a state that is like California in many ways, did not also pass a lifeline law. In order to answer that question, we need to take a closer look at differences in the regulatory environments of the two state commissions and at the goals of their principal regulators.

In terms of the regulatory environment, lifeline's failure in New York could have been due to the opposition of important groups or to the inability of pro-lifeline groups to generate needed support, or to both. New York, like California, is a hotbed of activity by left-liberal groups. As was the case in California, the 1973 oil embargo provided these groups with a useful organizing issue. Indeed, the impact of the embargo was even greater on a utility like Con Ed, which at the time generated 75 per cent of its electricity by burning imported oil, than on a utility like PG & E, which could rely to a greater extent on other fuel sources, including cheap hydro power. Moreover, the rates in New York were much higher than they were in California. In 1974 a San Franciscan's electricity bill amounted to $8.91 per month for 300 kwh; by contrast, a New Yorker was paying $24.22—the highest rates prevailing among the nation's twenty-five largest cities, and almost three times as much as those in San Francisco.[1] With such high rates one would expect that anti-utility feeling was running high in New York in the mid-1970's, and it was. Nevertheless, a comparison of the two states suggests that lifeline failed in New York primarily because consumer groups failed to channel this hostility into support for the lifeline concept with as much skill as did their California counterparts.

There are several reasons for this. First, although the New York Commission under Kahn's predecessor, Joseph Swidler, had its share of public relations problems, it did not appear to be as blatantly pro-business as did the California Public Utilities Commission at the time of Governor Reagan. Swidler had the reputation of being a professional, fair-minded regulator, who had demonstrated his independence throughout a lifetime of service on the Tennessee Valley Authority and, under Democratic presidents, on the Federal Power Commission. The Reagan CPUC, on the other hand, was perceived even by business executives as reckless and "too much in favor of business for business' own good." Mike Miller and Tim Sampson found that the California Commission was a useful "devil" in organizing the Citizen's Action League. Perhaps because the New York Commission did not appear to be as "evil" as its California counterpart, organizing effective groups was more difficult there.

Second, Kahn, another professional, replaced Swidler as chairman of the commission in June 1974, before the gubernatorial race in New York had heated up. Kahn's credentials, plus the reputation of Swidler before him, safeguarded the commission from some attacks to which it might otherwise have been subject. In California, by contrast, the commission became a lightning rod for opposition from left-liberal political entrepreneurs who made the commission a major issue in that state's gubernatorial campaign. One of Governor Jerry Brown's campaign promises was to "open up" the CPUC and "make it more responsive"—a pledge he followed up on by appointing Leonard Ross to the commission.

Third, in California the CPUC and the state's largest electric utility, PG & E, both had their headquarters in San Francisco, the media center of northern California and the home of many of the poor and older people who formed the base of the CAL organization. This accident of location lowered the cost and increased the returns of CAL's entrepreneurship. In New York, the location of the NYSPSC in Albany probably insulated it from some consumer group activity that it would have seen had its location been in New York City.

Finally, the New York lifeline supporters failed to create the coalition of environmental groups, consumer groups, and labor that was formed in California. The California coalition organized around "inverted rates" which promised conservation for the environmentalists and low prices for the residential user and which neutralized the opposition of labor. Although it began to weaken toward the end of its campaign, the California coalition lasted long enough for the legislature to stamp its proposal into law. No comparable coalition was formed in New York.

In addition to these environmental explanations, the personalities of Leonard Ross and Alfred Kahn made an important difference. Despite the similarity in their backgrounds, Kahn and Ross simply wanted different things. Kahn wanted to use the power of the state to improve the allocation of economic resources. To the extent that Kahn was motivated by personal concerns, it was to enhance his professional reputation—which reinforced his desire to create policy that increased economic efficiency. What is more, Kahn felt that in the give-and-take of regulatory politics, lifeline rates could threaten his efforts to gain organizational and political commitment for peak load rates, despite theoretical arguments that attempted to rationalize the combination of the two rates. (This, of course, is the same concern—but with the reverse effect—that motivated peak load pricing advocates in California, and is evidence of the greater relative power of California's consumer groups on the issue of rate reform.)

Leonard Ross, while not unappreciative of the importance of economic efficiency, saw other priorities. In the lifeline issue, he saw an opportunity to refashion the image of the California Public Utility Commission and to create a political reputation for himself. To the extent that he was motivated by other than personal political interests, Ross could justify lifeline publicly as a conservation measure (although his training in economics was too good for him not to have been aware of the flaws in this argument), and he could take economic solace in the Ramsey rule of inverse price elasticity. Perhaps Ross also felt, like the British socialist Peter Townsend whose opposition to the use of a "means" test in restricting social services to the poor had been cited by Ross in *Retreat From Riches*, that a universal lifeline rate was grounded in humane principle as well as in political calculation.[2] In sum, support for lifeline rates fit both the political and purposive motivations of Leonard Ross, just as opposition to lifeline rates fit both the political and purposive motivations of Alfred Kahn. In the case of lifeline (unlike that of regulatory lag) these motives mattered.

Asked to evaluate the various reasons why lifeline failed to get legislative backing in New York, one liberal New York state senator who was active in public utility issues during the 1970's, offered the following opinion:

> [*When the lifeline issue was brought before the legislature in 1975, the NYSPSC*] *was conducting its generic hearing and most people weren't paying close attention to what Fred was doing.*

The first lifeline bill didn't get out of committee because the Commission came on and opposed it. I think Fred had a lot to do with why lifeline wasn't passed. In about March 1975, he came before the legislature to testify on lifeline. People were very impressed. Afterwards, I was interviewed as to my reaction and I said that I found him to be knowledgeable and fascinating to listen to. Fred saw me say that on TV, and I think he thought here is someone who might make a useful ally. He called and asked that we have lunch and we did. We spent about three hours together and then for about four months thereafter he would send me things daily—speeches, notes.

I think there are really a couple of other reasons why it didn't pass. One, you couldn't support lifeline on equity grounds because unless the blocks were chosen very carefully, lots of people you were going to try to help were going to get screwed. Downstate, we have a lot of people in energy inefficient housing, and they would be hurt by higher rates for high use customers. Then, too, since these people were on master system meters, if you reduced their rates, the landlord might just pocket the difference.

Two, I think it's very important that California was experiencing an economic boom [at the time]. In New York, we had lost hundreds of thousands of jobs over the past decade and there was real concern that restructuring the rates might just tip the balance.

Finally, our consumer groups are split in New York. There wasn't the organization that occurred in California or the sophistication. And there were other issues occupying the attention of the groups and the media. At the time, if you asked the question what was occupying the attention of the press, it was nursing homes. You had to ask yourself what's more important for the elderly, nursing homes or electricity?

Different Contexts, Different Tasks

If in accounting for the differences in regulatory outcomes in New York and California we must keep in mind that Alfred Kahn and Leonard Ross faced different political contexts, then we must also recognize that Kahn's priority commitment to peak load pricing presented him with bureaucratic imperatives that did not constrain Ross in the same way, despite the similarity of the two agencies that each headed. Ross's political entrepreneurship and his short time horizon freed him from worrying about the need to induce cooperative behavior from the staff beyond the minimal amount of effort that one could expect from civil servants and those few staff members who happened to agree with him. He was not fundamentally interested in making California the leader in designing economically efficient electric rates, although he did support their adoption. On the other hand, efficient electric rates were a central concern to Alfred Kahn. Kahn was in the process of establishing himself as the dean of American regulators; it was likely that at some point he would be going to Washington, and he had already devoted a lifetime of scholarly analysis to regulatory matters. He wanted and needed organizational commitment to his reform proposals so that the impact of his initiatives would continue to be felt long after he left the New York State Public Service Commission. The bureaucratic problem that Kahn faced was not simply that of inducing subordinates working within his agency to behave as he would were he in their place; it was more. Kahn had to convince the members of his staff to continue to apply and develop the principles that he had taught them long after he left, under economic and technological conditions that he knew would be different from those prevailing at the time of his tenure as chairman.

It was an extraordinary intellectual and managerial task that Kahn took upon himself. To secure organizational commitment both for the short term and for his longer term plan, he would have to pay careful attention to the professional and personal incentives that motivated his staff, and he would have to make good use of the agency's informal network of contacts. He did both. Informal contacts within the staff were continually nourished. Communication (both up and down the agency's hierarchy) was fostered by such casual affairs as noon-time swimming breaks to which Kahn invited both division heads and freshman attorneys. But if Kahn's informal and collegial manner is what won the hearts of his professional staff, it was his intellectual brilliance, dazzlingly displayed in impromptu seminars and in generic hearings, that won their minds.

Some measure of Kahn's success in this effort is contained in the enthusiastic comments of one young rate engineer in the power division, who by his own admission, knew very little about peak load pricing before Kahn joined the commission. Looking out his office window at the sweeping architectural design of Albany's marbled Empire State Plaza sometime after Kahn had left for Washington and the CAB, he was moved to say, "You know, when I think about the demand management techniques that we are attempting to put into effect here in New York, I start to think that perhaps, just maybe, they could help us avoid building even one nuclear power plant. And if we could do that we'd save $1 billion, which would free up enough resources to build another plaza just like this. The energy crisis is going to be the major problem of our generation. Rate design is right on top of it, and this is New York! You know, lots of people are talking about rate design now. I tell you I get excited to hear President Carter talking about time-of-day pricing."

This fervor for rate reform was noticeably absent in California, although even there a number of staff engineers were becoming convinced by the arguments of the economists. Certainly, staff members evidenced very little personal loyalty for the commission majority. In a 1977 interview, one of the CPUC attorneys explained that this actually had very little to do with the substance of rate reform:[3]

Question: When we talked about the morale of the commission staff last year, you mentioned that it wasn't very good. Do you recall that? Can you contrast it with now?

Staff Attorney: Yes, I do recall saying that, and if anything the morale is worse today. People are around here on the job and when 5:00 comes—boom they're gone and that's it.

Question: Oh, really? That's interesting. You know when I was in New York, I was interested in seeing the staff reaction to Chairman Kahn. Here was an economist. . . .

Staff Attorney: Kahn? The man who became chairman of the CAB?

Question: Yes. Here was a man who brought with him a firm set of notions of what he wanted to do. He attempted to incorporate principles of economics into his ratemaking: time-of-day pricing in electricity and marginal cost in telephone. These principles differed from the way engineers were doing things. He also instituted more informal hearings—swearing people into panels and then running them as seminars. You might have thought on the face of it that he'd get a lot of opposition from engineers, attorneys, and administrative law judges. But the morale, surprisingly, was very good. Not only was there respect for the man, there was affection.

Staff Attorney: Well, that's because he was sincere. See, he could have gotten a lot of support around here, too. But our commissioners just aren't sincere.

Question: Well, I'm trying to understand why Kahn was successful at gaining staff support and this commission is not.

Staff Attorney: Around here it's all political. The staff—the professional staff (and this is, of course, just my opinion)—would go along with a reasoned approach that was thoughtful and professional. We're upset because all the commission does is issue a press release and then when the public attention dies down they lose interest and go on to something else.

Question: Then maybe it's not so much the content of the policy as the approach? Because in New York the commission is going in similar directions as California except on lifeline.

    Staff Attorney: That's right. Kahn's appeal sounds like it was professional. Here it's just plain politics.

The Dual Nature of Regulatory Agencies

The striking difference in the personal styles of Alfred Kahn and Leonard Ross serves to drive home a point. When the primary task facing a regulator is a complex, technical one (what we have called a planning task) the regulator will be compelled to respond to internal organizational needs. Under such circumstances the bureaucratic nature of a regulatory agency asserts itself. But regulatory agencies have a dual nature. When the principal task faced by a regulator is instead one of making a single choice between (among) competing values—such as was the case with the lifeline decision—regulators can ignore, if they choose, the maintenance needs of their agency. It is then that the political nature of a regulatory agency becomes paramount.

    The very fact that the regulatory environment is a dynamic one in which both tasks and external constraints change and affect each other, however, poses a dilemma for the regulator, especially for a political entrepreneur who sees his tenure as a highly visible head of a regulatory agency as a stepping stone to higher office. A political appointee can ignore his staff for only so long in using his regulatory decision-making power to forge a coalition of support within the political market. At some point, first the staff, and then the public through the staff, will begin to question the legitimacy of using the regulatory process for strictly political purposes. To establish credibility, or reestablish it once it has been jeopardized, the political entrepreneur will necessarily have to defend his agency's procedures for receiving expert testimony and for arriving at decisions through detached, scientific deliberation. To be convincing in this effort, he will need the cooperation of the staff, and this will subject him to internal constraints that will limit the scope of his entrepreneurship. To avoid these internal constraints, a political entrepreneur has to move quickly. Either he must accomplish his purposive and political objectives early in his tenure before the staff becomes aroused, or he must recognize that his designs have been frustrated (for whatever reason) and leave the agency before the staff has a chance to strike back and discredit him.

This is a point that did not escape Leonard Ross. Although the staff of the California Public Utilities Commission became demoralized and bitter about his "regulation by press release style," organizational strain did not hinder Ross in the short run from using the lifeline issue to generate favorable publicity and public recognition. But Ross recognized the limits to the degree to which he could use the CPUC as a political organ. He avoided bureaucratic constraints only by cutting short his tenure in office. (He left the CPUC in early 1977 after less than two years to serve as special assistant to Richard N. Cooper, Undersecretary of State for Economic Affairs.) Had he stayed longer as commissioner, Ross would surely have bumped up against staff-imposed road blocks; it is possible that by the time he left the agency these limits were already operative. The general observation is that even in agencies that have been politicized by policy entrepreneurs, organizational imperatives cannot be ignored forever. At some point the professional staff of an agency will seek to reassert its control over the regulatory process. Agency careerists will find this opportunity in the next planning task in which an offensive regulatory executive has an important stake. It is then that the analyst will need to look for bureaucratic determinants in explaining regulatory outcomes.[4]

Regulatory Bureaucracy

It is not, as George Stigler has maintained, a hollow generalization to conclude that agency staffs "exert a major influence on the scope and direction of policies."[5] Indeed, for too long, a systematic understanding of how bureaucratic imperatives shape and affect regulatory outcomes has been impeded by economic theories of regulation that have treated regulatory agencies essentially as if they were black boxes, bent on maximizing along a single dimension and taking all of their cues from the external environment. Even when Stigler's theory is generalized to allow for the possibility of capture or purchase by non-business groups, it still relies on the dubious causal chain that regulatory outcomes are determined by political appointees who are disciplined and controlled by the few political parties which sell the coercive power of the state.

There are two tenuous links in this model. For one, the model ignores two highly significant recent developments in American politics: the rise of entrepreneurial politics and the decline of political parties. These developments are, I suspect, linked. Political entrepreneurs develop public reputations and power by functioning within the political process and using governmental institutions without the assistance or control of political parties and machines. As long as politicians were dependent upon parties for money and organizational efforts in waging campaigns and getting out the vote, parties could effectively discipline them. Now politicians do not rely on parties as the major source of campaign finances; nor are they as dependent upon parties for organization and party identification. Television has provided individual politicians with a mechanism for reaching large numbers of voters directly, a development which has both short-circuited the informational role of parties and broadened the political market. Unable to help, parties can no longer hurt.

These are structural changes in the political market that Stigler's duopoly or oligopoly theory does not comprehend. It only slightly exaggerates the point to say that Congress now consists of 535 parties rather than two or four.[6] The chief significance of these developments is not only that purchase of state power is now more complicated; it is that the sanctity of contracts between regulators and client groups is no longer (if it ever was) inviolate. A new type of politics, one that was not anticipated by the economic theories of regulation, is increasingly the norm in situations where entrepreneurial politics plays a role. In such instances, maximal squawk has replaced minimal squawk as a description of the basic behavior of regulatory commissions.

The second problem with Stigler's causal chain is that it makes the implicit assumption that one can ignore internal, bureaucratic constraints. Once an agreement has been struck between a client group and a regulator, the regulator will take whatever steps are necessary to translate the agreement into the appropriate regulatory policy. But what if the policy offends the staff? Does it matter? Can the staff force the regulator to amend his policy? Can it force him to abandon the regulatory initiative altogether? If so, under what circumstances? It is here that the economic theory of regulation fails us most, for it is silent on these questions. It certainly does not explain the choice of policies that John Vukasin used to reduce regulatory lag, or why he was unsuccessful in his program of staff rotations and procedural changes. Nor can it account for the changes in the relationship between the California PUC and electric utilities that followed in the wake of Vukasin's term.

The analogy of a maximizing firm is appropriate only in anticipating the forces leading to agency capture and in understanding regulatory behavior under conditions of capture. It is useful then because internal constraints are inoperative and because external constraints (by a single source) are totally binding. In this case it can be assumed that the agency tries to maximize along a single dimension, and that is to do whatever pleases the client that has become its master. In any other situation the simple maximizing assumption is not helpful, either because there are no binding external constraints (the Type III entrepreneurial case), because there are competing external and internal constraints (the Type II conflict-minimizing case) or because the constraints are principally internal (the Type IV bureaucratic case).

Extending Theory

The theoretical component of this book is designed to elucidate the circumstances under which bureaucratic imperatives are significant determinants of regulatory behavior and performance. We have seen that these constraints are sometimes important, and sometimes not. Content has been added to the generalization that regulatory staffs matter by emphasizing that *what* a regulatory executive attempts to do is as significant (or more so) in many cases as *why* he is trying to do it. In choosing this path, attention has been focused on the interplay between the nature of agency tasks and forces in the agency's environment. This does not provide a complete theory of regulation, however. While suggestive, it cannot be said to be definitive.

To be complete, the theory of regulation must help us understand more about the dual nature of regulatory agencies as political and bureaucratic organizations. Further conceptual development is needed to predict which client groups will benefit most from regulation. This is the venerable *cui bono* question of political analysis, and in terms of this model, it amounts to a development of the theory of supply. Extending the theory in this regard would be no small undertaking. It would involve at least the complexity of theories that attempt to predict which firms will perform best on the stock market. Furthermore, both more empirical work and more conceptual refinement is needed in understanding the dynamics of regulatory bureaucracy. How do bureaucratic imperatives arise and how do they vary across agencies? This study has focused on the incentive system and informal organization of agencies. Both of these concepts need elaboration and further study in the regulatory context, as do other factors that affect the ability of an agency to define its mission and accomplish its tasks. We need to know more about the factors which animate the relationships among career bureaucrats, regulatory professionals, and political appointees, and we need to know more about how organizational roles shape and affect the members of a regulatory bureaucracy.[7]

A Now Familiar Refrain

Taken together, the California and New York experiences with electric utility rate reform illustrate a simple, yet often overlooked fact. Far more than is commonly recognized, regulatory politics depends upon dynamic factors in the economic and technological environment of regulation. Most studies—even those that purport to be based on "life-cycle theories"—offer a static snapshot of an agency coping with "industry" but give little sense of the way in which technology and market forces alter that industry and thus create new problems for the agency.

In the case of electric utility regulation by the states, major controversies involving "cost-of-service" ratemaking procedures were resolved by the end of World War II. Commissions were charged with the obligation of assuring "just and reasonable" rates, good-quality service, and the avoidance of "undue discrimination." While these terms remain vague to the layman, they are terms of art, given content by a long history of judicial interpretation. In practice they came to mean that the task of regulators was to protect consumers against excessive rates and to protect investors against loss of property, while utilities were granted the right to structure rates in a way that would yield the commission-determined "revenue requirement." As long as technological opportunities promised lower costs through increased usage, utilities were given a free hand to adopt rate structures that promoted consumption. Large users clearly benefited from this policy, but so did consumers, investors, and, as a result, regulators.

When in the late 1960's and early 1970's these technological opportunities were largely exploited, the practice of allowing utilities to structure their own rates ceased to be noncontroversial. As the cost of construction and debt escalated, the savings achieved through larger-scale plant diminished. For utilities that suffered from a deteriorating load factor, it no longer was economical to encourage high-volume usage—especially if taken "on-peak." These trends reached crisis proportions following the OPEC oil embargo of October 1973. Fuel cost adjustment clauses helped utilities raise large amounts of working capital, but these devices did not solve the utilities' longer term production problems.

These technological and economic changes completely altered the political context of state utility commissions. Enormous rate increases stimulated the organizational efforts of angry residential consumers, who rediscovered the regulatory process at the same time environmentalists began pressuring both utilities and regulators to reexamine the industry's progrowth ethos. Where once had existed a sort of tacit coalition between large and small users, there was now open hostility. Instead of having the happy task of presiding over a process that seemed to bestow benefits upon everyone, regulators in the 1970's were faced with the unpleasant duty of allocating misery among rival and intensely vocal groups.

The regulatory response to these new requirements has not been predetermined. There is an element of choice in regulatory behavior. How this choice is exercised depends on a number of factors. One is accidental: who happens to be in charge and when. Alfred Kahn's efforts on the New York Public Service Commission illustrate the impact one person can have on the regulatory process. Were it not for Kahn, New York would probably not have emerged as the leading state on the matter of peak load pricing, but just as important, had Kahn been chairman in 1964 instead of 1974, he probably would not have been able to move the commission and the utilities to adopt marginal-cost-based rates. Another factor explaining regulatory behavior is more patterned: the emergence of a new political career for policy entrepreneurs in the regulatory field. The positions of Ross, Batinovich, and Holmes in the California lifeline case were all motivated by what they considered to be opportunities for political advancement created by the controversy surrounding the commission.

Certainly in California and New York, commissions have not responded to the changed industrial and political environment as if they were "captured" by the utilities or their largest users. Old notions of regulatory "capture" were based on the assumption—perhaps valid in the case of electric utilities during the 1950's and 1960's—that industry exercised exclusive control over the relevant incentives for regulators. Now industry obviously does not. Either there are no external incentives (Kahn, at the peak of his career, was much more concerned that his actions conformed to the norms of his profession than whether they happened to please some special interest), or the incentives are under the shared control of industry, the agency's bureaucratic staff, the media, and the political process. The experience of California utilities at the time Leonard Ross and his associates directed the Public Utility Commission suggests that industry's share of control may even approach zero.

Endnotes

1. Public Utilities Commission, State of California, *Annual Report 1973–1974 Fiscal Year* (San Francisco, 1974), p. 80.
2. In a chapter from his book entitled "Sharing the Wealth: Income Redistribution as an Alternative to Growth," Ross noted the opposition of Britain's Labour left to the "means" test. In describing the basis of this opposition, Ross and his co-author Peter Passell wrote:

"The left-wing obsession with the means test was founded on both humane principle and political calculation. The means test, Peter Townsend recently wrote in a Fabian pamphlet, 'fosters hierarchical relationships of superiority and inferiority in society, diminishes rather than enhances the status of the poor, and has the effect of widening rather than reducing social inequalities.' . . . If, on the contrary, social programs were extended to all, the groundwork would be laid for that 'complex reconstruction of the systems of reward in society' which would truly spell socialism."

See Peter Passell and Leonard Ross, *The Retreat from Riches: Affluence and Its Enemies* (New York: The Viking Press, 1971), p. 83.
3. This interview was a follow-up of an interview held one year earlier with the same California Public Utilities Commission staff member. The object of the interview was to gain information useful in comparing and contrasting the regulatory styles of Commissioner Ross and Chairman Kahn. It is representative of the views expressed in a number of other such interviews.
4. Leonard Ross was unable (at least by late 1980) to make use of his CPUC reputation in moving up the elective ladder in California politics. But his failure

to do so does not detract from the basic observation that his performance during his tenure as a member of the California Public Utilities Commission needs to be viewed as entrepreneurial to be understood.

Political entrepreneurship by the heads of regulatory agencies needs to be more systematically studied. I am not aware of any study that documents the use of regulatory positions by politicians seeking election to higher office, but I can think of at least two recent examples. In 1974 John Durkin used the reputation he developed as a consumer advocate while New Hampshire insurance commissioner to win election to the United States Senate. In 1980 Paula Hawkins, who first developed a public reputation as chairman of the Florida Public Service Commission (and was an advocate of inverted rates for electricity), followed the same path.

5. George J. Stigler, "Trying to Understand the Regulatory Leviathan," *The Wall Street Journal*, August 1, 1980.

6. On the concept of Congress as a four-party system *see* James MacGregor Burns and Jack Walter Peltason, *Government by the People: The Dynamics of American National Government*, seventh edition (Englewood Cliffs: Prentice-Hall, 1969), pp. 291–293.

7. There have been efforts in this direction. *See* James Q. Wilson, ed., *The Politics of Regulation* (New York: Basic Books, 1980). The chapter by Wilson classifies agency members according to their motives in seeking to expound the bureaucratic determinants of regulatory behavior (pp. 372–382). The chapters by Suzanne Weaver and Robert Katzmann are particularly interesting in describing and evaluating inter-professional rivalry between economists and lawyers in the Department of Justice and in the FTC on the question of whom to prosecute for antitrust violations. *See also* Suzanne Weaver, *Decision to Prosecute: Organization and Public Policy in the Antitrust Division* (Cambridge, Mass.: MIT Press, 1977) and Robert A. Katzmann, *Regulatory Bureaucracy: The Federal Trade Commission and Antitrust Policy* (Cambridge, Mass.: MIT Press, 1980).

INDEX